T0418501

The Critical Ihde

SUNY series in American Philosophy and Cultural Thought

Randall E. Auxier and John R. Shook, editors

The Critical Ihde

Edited by

Robert Rosenberger

Published by State University of New York Press, Albany

© 2023 State University of New York

All rights reserved

Printed in the United States of America

For information, contact State University of New York Press, Albany, NY
www.sunypress.edu

Library of Congress Cataloging-in-Publication Data

Name: Rosenberger, Robert, editor.
Title: The critical Ihde / Robert Rosenberger.
Description: Albany : State University of New York Press, [2023] | SUNY series in American Philosophy and Cultural Thought | Includes bibliographical references and index.
Identifiers: ISBN 9781438492612 (hardcover : alk. paper) | ISBN 9781438492629 (ebook)
Further information is available at the Library of Congress.

10 9 8 7 6 5 4 3 2 1

Contents

PART 1
FROM PHENOMENOLOGY
TO POSTPHENOMENOLOGY

PART 2
THE PHENOMENOLOGY OF TECHNOLOGY

PART 3
THE PHENOMENOLOGY OF SCIENCE

Illustrations

Acknowledgments

All chapters are reprinted from previously published material by Don Ihde. These include, in chapter-by-chapter order, the following:

- "The Auditory Dimension" and "The Shapes of Sound," chapters 4 and 5 of *Listening and Voice: Phenomenologies of Sound*, 2nd ed. (Albany: State University of New York Press, 2007), 49–71.

- "Illusions and Multi-Stable Phenomena: A Phenomenological Deconstruction" and "Variations Upon Deconstruction: Possibilities and Topography," chapters 4 and 5 of *Experimental Phenomenology: An Introduction*, 2nd printing (Putnam, 1977; Albany: State University of New York Press, 1986), 67–90.

- *On Non-foundational Phenomenology*, ed. Seth Chaiklin, Fenomenografiska notiser 3 (Gothenburg: Institutionen för pedagogik, Göteborgs universitet, 1986).

- "Response to Rorty, or, Is Phenomenology Edifying?," chapter 9 of *Consequences of Phenomenology* (Albany: State University of New York Press, 1986), 181–98.

- "What Is Postphenomenology?," chapter 1 of *Postphenomenology and Technoscience: The Peking University Lectures* (Albany: State University of New York Press, 2009), 5–23.

- "Program One: A Phenomenology of Technics," chapter 5 of *Technology and the Lifeworld* (Bloomington: Indiana University Press, 1990), 72–123.

- "Embodying Hearing Devices: Digitalization," chapter 22 of *Listening and Voice*, 2nd ed. (Albany: State University of New York Press, 2007), 243–50.

- "Deromanticizing Heidegger," chapter 8 of *Postphenomenology: Essays in the Postmodern Context* (Evanston, IL: Northwestern University Press, 1993), 103–15.

- "Technology and Cultural Variations," chapter 5 of *Consequences of Phenomenology* (Albany: State University of New York Press, 1986), 116–36.

- "Technology and Prognostic Predicaments," *AI and Society* 13 (1999): 44–51.

- "Husserl's Galileo Needed a Telescope!," *Philosophy and Technology* 24 (2011): 69–82.

- "The Historical-Ontological Priority of Technology over Science," chapter 2 of *Existential Technics* (Albany: State University of New York Press, 1983), 25–46.

- "Art Precedes Science: Or Did the Camera Obscura Invent Modern Science?" In *Instruments in Art and Science*, ed. Helmar Schramm, Ludger Schwarte, and Jan Lazardzig (Berlin: de Gruyter, 2008), 383–93.

- "Scientific Visualism" and "Technoconstruction," chapters 12 and 13 of *Expanding Hermeneutics: Visualism in Science* (Evanston, IL: Northwestern University Press, 1998), 151–89.

Introduction

Robert Rosenberger

The time has come to assemble a collection of critical works by American phenomenologist Don Ihde. As Distinguished Professor of Philosophy Emeritus at Stony Brook University, Ihde continues to be a leading figure in thought on the place of technology in our lives. His work has inspired countless scholars. His ideas have been applied by others to an extensive range of topics, empirical studies, design projects, artworks, and philosophical problems. And his own extended case studies have provided a model for in-depth philosophical engagement with the specific details of our contemporary technological situation. For anyone working on issues of technology, Ihde's ideas have the potential to provide useful tools for articulating the ways that the concrete materiality of our devices gives shape to human experience.

Ihde's body of work has had an expansively interdisciplinary impact. His writings are well known within the discipline of philosophy in both Continental philosophical circles and in the philosophy of science, and he has influenced multiple other fields, including media studies, human-computer interaction, sound studies, design, and many others. Without exaggeration, he is the "go-to" phenomenologist in both the fields of philosophy of technology and science and technology studies (STS).

The quantity of academic output over the course of Ihde's career has been immense. This has included the production of 32 books, including 25 monographs of various types (Ihde 1971, 1973, 1976, 1977, 1979, 1983, 1986a, 1986b, 1990, 1991, 1993a, 1993b, 1998, 2002, 2007, 2008b, 2008c, 2009, 2010a, 2010b, 2012, 2015, 2016, 2019, 2021) and 7 edited and coedited volumes (Zaner and Ihde 1973; Ihde 1974; Ihde and Zaner

1975, 1977; Silverman and Ihde 1985; Ihde and Silverman 1985; Ihde and Selinger 2003).[1] His work has been the subject of two festschrifts (Selinger 2006; Miller and Shew 2020), as well as numerous special issues of academic journals, including ones that contrast his work with other figures such as Andrew Feenberg, Lambros Malafouris, and Bernard Steigler.[2] In addition, Ihde can claim a number of "firsts": the first to write a dissertation in English on the work of Paul Ricoeur, the first book on auditory phenomenology (*Listening and Voice*, 1976), and the first English work in the philosophy of technology (*Technics and Praxis*, 1979). And he has been the recipient of multiple awards, including a Lifetime Achievement Award by the Society for Philosophy and Technology, a Walter J. Ong Lifetime Scholar Award from the Media Ecology Association, a Golden Euridici Award from the International Biophilosophy Forum, and induction as an American Association for the Advancement of Science (AAAS) Fellow.

In light of all of this, it seems that a collection of many of Ihde's most representative and influential works would be a handy thing to have. When approached by SUNY Press for this project, I was happy to take on the task. As someone who has utilized Ihde's ideas in my own studies for many years, and as someone who has operated in the middle of the burgeoning collective of scholars that has sprouted up around his work, I have a good sense of at least some of the parts of Ihde's thinking that have proven useful for many people. In what follows here in *The Critical Ihde*, we have collected an assortment of Ihde's writings that cover many of his varied contributions. We have here several of his classic and most influential pieces. We have some works from his corpus that contain examples of his most useful and original ideas, cases, and arguments. And we have several pieces that I consider to be underrecognized gems. Surely some Ihde aficionados will disagree with what I have included here in this collection, or, more, what I failed to include. In any case, I do believe this collection to be a strong introduction to Ihde's thinking, and a valuable resource for those that make use of his ideas.

One of Ihde's central contributions is the development of the "postphenomenological" philosophical perspective. This term has come to mean many things. It is a moniker for Ihde's own philosophy generally, the totality of ideas he has developed and the things he has written. But it also refers to a particular framework of philosophical concepts and commitments that have come out of his work, and the movement of scholars building on these ideas.

"Postphenomenology," as Ihde has developed it, brings together insights from the phenomenological tradition with the philosophical commitments of American pragmatism and other postmodern perspectives. These ideas are crafted into a practical framework of concepts for describing human relationships with technology. And this is one reason for Ihde's wide interdisciplinary appeal: his ideas are useful. They provide a practical toolbox of concepts that can be applied to the study of human-technology relations. They can be taken up by others for all kinds of analyses of technology. And they can be integrated into others' empirical methodologies. This perspective offers an enlightening way to approach a number of the central themes in work on technology, from the implications of technology's concrete materiality, to the specifics of our bodily-perceptual and human-interpretive relations to our devices, as well as to the situated and coconstituted nature of these relations. Ihde himself uses these ideas as part of decades-long case studies into laboratory instrumentation, ancient human technological practices, medical imaging, archery, hearing aids, writing technologies, nautical navigation techniques, musical instruments, and many other things.

A central aspect of Ihde's perspective is the suggestion that technologies should be conceived as "mediators" of human experience. Technologies are not just another category of things out in the world that we might perceive or act upon. They are a transformative means by which our perceptions and actions happen. They enable us to perceive and act differently. In this way, neither do technologies entirely determine what we perceive or how we act, nor do they function as innocent instruments that merely facilitate those perceptions or actions. As Ihde puts it, "There are no neutral technologies, or positively put, all technologies are non-neutral" (1993b, 34). Much of Ihde's career is thus occupied with following out the implications of the transformative and nonneutral nature of human-technology relations, exploring the concrete tradeoffs of design, the micro- and macroperceptual levels of mediation, and the material trajectories that guide scientific and technological development.

The concepts coming out of postphenomenology excel at the deep description of human experiences with technology. But they also specialize in capturing the variability of these experiences. Ihde is adamant that human-technology relations are never merely one type of thing. They never fit into only one category, even broadly speaking. Different possible relationships can always be identified. And the same technology can always be used in multiple ways, develop along different lines, and fit differently into different cultures. As Ihde writes, "Technological culture, I have been arguing, is not simply one thing. Neither is it uniform nor has its progression across the globe attained either what its detractors fear or what its proponents hope"

(1990, 150–51). This dual specialization of postphenomenology—at both the deep description of technological experience and also the recognition of its variability—is present within many of the main ideas of this perspective.

For example, this can be seen in what might be Ihde's most influential set of concepts: his quartet of human-technology relationships, that is, what he calls "embodiment relations," "hermeneutic relations," "alterity relations," and "background relations" (e.g., 1990, 72–112; 1999, 42–44). These ideas (which we'll see articulated and used over and over through this book, esp. ch. 6) and their associated concepts help to describe the ways a user engages the world through technology. At the same time, as itself a list (and an open-ended one at that, and one that has been expanded on by others; see esp. Verbeek 2011), the identification of these different possibilities for engagement itself serves to emphasize one aspect of the variability of our relationships with technology.

A second central idea Ihde has developed for capturing both the specificity and the variability possible for human-technology relations is the notion of "multistability." Originally crafted as a way to describe the multiplicity possible for human visual perception (see ch. 2), this idea helps to capture the ways that technologies never reduce to only one meaning or usage. At the same time, this idea also recognizes that a given device cannot simply mean anything, nor can it be used for just any purpose; human-technology relations are limited to particular "stabilities" (or "variations"). Ihde often uses this idea as a tool for criticizing other theoretical accounts that he judges to be essentializing or overgeneralizing, that is, accounts that allegedly fail to recognize technology's multistability. (This idea shows up over and over throughout this book, e.g., chs. 2, 3, 4, 9, and 10.)

One crucial consequence of this account of technological mediation is that we cannot think of ourselves or the world as separate and pregiven entities. We become who we are, and the world is revealed for what it is, through the mediation of technology. This commits postphenomenology to what is sometimes called a "relational" (or sometimes "interrelational") ontology, an understanding of the basic nature of things as fundamentally constituted through their relations to other things. As Ihde writes, "As can be seen, in each set of human-technology relations, the model is that of an interrelational ontology. This style of ontology carries with it a number of implications, including the one that there is a coconstitution of humans and their technologies" (2009, 44). This conception of ontology operates in tune with postphenomenology's philosophical commitments to the American pragmatist perspective, and it puts Ihde in the company of other contemporary postmodern thinkers on technology (e.g., Donna Haraway and Bruno Latour). It also has set up postphenomenologists to explore the ways that technological mediation coconstitutes our moral and political situation (e.g., Verbeek 2011; Rosenberger 2017).

In recent years the term "postphenomenology" has come to refer not only to Ihde's own corpus of books, ideas, and positions on various topics, but also to the work of a growing international and interdisciplinary collective of scholars who are advancing and applying these ideas. (And it can sometimes be difficult to disentangle these two meanings.) These scholars include Dutch phenomenologist of technology Peter-Paul Verbeek and Danish anthropologist of education Cathrine Hasse, among many others (e.g., Verbeek 2011; Friis and Crease, 2015; Rosenberger and Verbeek 2015; Wellner 2015; Irwin 2017; Rosenberger 2017; Hasse 2020; de Boer 2021). This work includes as well the book series "Postphenomenology and the Philosophy of Technology" with Lexington Books, which now comprises a stack of volumes on a variety of topics (including an assortment of perspectives from within the fields of philosophy of technology and STS), and which contains an ongoing line of books with a title structure of "Postphenomenology and X" on topics such as media, architecture, methodology, and imaging (e.g., Van Den Eede, Irwin, and Wellner 2017; Aagaard et al. 2018; Botin and Hyams 2021; Fried and Rosenberger 2021). Panels on postphenomenological research have become a reliable fixture of a number of international conferences, often with Ihde himself as the anchor presenter. The bustling postphenomenological research movement stands as another testament to the impact of Ihde's work.

Ihde's distinctive style is another feature of his writing reflected in the selections included in this book. In addition to the expected textual analysis of the writings of canonical figures from the history of philosophy, we see an excited inquisitiveness regarding the details of technology in the various contexts of everyday life, from the home to the laboratory. Ihde's playful philosophical engagement with the world can be liberating, especially for those of us with empirical and interdisciplinary leanings. In addition to the expected philosophical texts, Ihde pulls from world history, biological and cultural anthropology, the daily news, and the latest scientific journal articles across a number of fields. Phenomenology, as practiced by Ihde, is thus not merely some corner of philosophical study concerned mainly with textual exegeses of canonical figures or the fiddling with abstract and esoteric philosophical quandaries; it is something that one *does*. "Let's do some phenomenology," he often says.

In addition, as readers of Ihde know well, he often pulls from his own life. Much of his work has an autobiographical flavor. Across his texts we learn about his life growing up on a farm in Kansas where he was educated in a one-room schoolhouse, life in places like New York City and his retreat in Vermont, his experiences sailing and painting, his worldwide travels exploring ancient ruins or high-tech virtual reality and robotics laboratories, and the details of his open-heart surgery, among many other anecdotes, observations, and personal notes. Thus, in addition to the philosophical contributions of his work, there is also a lively spirit to it, an animated curiosity, an ironic playfulness, and an enthusiastic fascination with the details of human lives, contemporary technologies, cutting-edge science, animal studies, and anthropological history. This style and spirit continues to inspire contemporary scholars.

The selections in this volume are arranged into three parts.

Part 1: From Phenomenology to Postphenomenology

The essays in part 1 provide key examples of Ihde's contributions to phenomenology proper, and to his development of his own pragmatism-infused "postphenomenological" perspective.

Chapter 1 represents one of Ihde's most influential and original phenomenological investigations: his phenomenology of sound. This work culminated in his 1976 book *Listening and Voice*, which explores the nature of auditory experience. Ihde explores the spatial and temporal dimensions of sound, and the ways auditory experience is a central part of our encounter with the world. He also considers how the experience of the special presence of other people can occur through auditory perception. Chapter 1, here entitled "Auditory Phenomenology," is comprised of selections from *Listening and Voice*.

Chapter 2, entitled "The Multistability of Perception," is a phenomenology of visual experience. Ihde's work on this topic has a special focus on the variability possible for human perception and the embodied hermeneutic dimensions of what we see. This work includes his exploration of the gestalt experience of visual illusions. And through these investigations he develops the notion of "multistability," an idea later applied as well to the philosophy of technology. This chapter is comprised of selections from his 1977 book *Experimental Phenomenology*.

The next three chapters concern Ihde's development of the "postphenomenological" perspective, and in particular his specific amalgamation of phenomenological insights with the philosophical commitments of contemporary American pragmatism. Chapter 3, here entitled "What Pragmatism Adds to Phenomenology," is comprised of an underappreciated gem from Ihde's corpus: his 1986 work *On Non-foundational Phenomenology*, reproduced in its entirety. The piece is a transcription of a lecture on the implications of Rortian and Foucaultian postmodern philosophy for phenomenology, including Ihde's responses to audience questions. Chapter 4, entitled "What Phenomenology Adds to Pragmatism," continues the development of these ideas, clarifying how phenomenology—properly understood, according to Ihde—not only survives the kinds of attacks that Richard Rorty has leveled against other philosophies, it actually provides distinctive ways forward. This chapter comes from "Response to Rorty, or, Is Phenomenology Edifying?," from his 1986 book *Consequences of Phenomenology*. In chapter 5, here entitled "What Is Postphenomenology?," Ihde pulls all of this together into an introduction to the postphenomenological perspective. It comes from the opening chapter of the same title from his 2009 book *Postphenomenology and Technoscience: The Peking University Lectures*. This includes a number what have become hallmark strategies in Ihde's project of defining and clarifying what postphenomenology is all about: the critique of Husserl, the integration of ideas from Dewey, and the utilization of visual illusions to help articulate the multistability of our relationships with technology.

Part 2: The Phenomenology of Technology

Part 2 of this book is on what is surely Ihde's most influential and useful set of ideas: his account of the human bodily-perceptual experience of technology. Postphenomenology is often understood as a perspective in the philosophy of technology, and these chapters explore the various concepts, ideas, and examples of Ihde's account.

Chapter 6, here entitled "Human-Technology Relations," includes what is probably Ihde's most influential individual set of ideas: his fourfold list of "existential" relations to technology. First introduced in his 1979 book, *Technics and Praxis*, and developed throughout the 1970s and 1980s, these ideas achieved what is often regarded as their definitive version in the selection included here from his 1990 book *Technology and the Lifeworld*.

It is possibly the most widely taken-up piece of writing in Ihde's entire corpus. This chapter includes detailed and technical articulations of the four different bodily engagements with technology that he identifies, that is, "embodiment relations," "hermeneutic relations," "alterity relations," and "background relations." It also includes, among other things, an account of "technological transparency," Ihde's practical description of Heideggerian withdraw, in both its embodied and hermeneutic dimensions.

Ihde's contributions to the phenomenology of sound continue into his philosophy of technology. He brings these insights to everything from electronic instrumentation to sonar tomography, and such explorations are featured across his work, but most centrally in his 2007 expanded second edition of *Listening and Voice* and his 2015 book *Acoustic Technics*. In chapter 7, here entitled "Auditory Technologies," we see an example of this work in the articulation of the experience of hearing aid technologies. Ihde is himself a longtime wearer of these devices, so he builds his account on years of personal experience.

Another central feature of Ihde's philosophy of technology has been his critique of the work of Martin Heidegger. Heidegger serves as both inspiration and foil for so much of Ihde's writing throughout his career, and these analyses culminate in his 2010 book *Heidegger's Technologies*. On the one hand, the fingerprints of Heidegger's thought can be lifted from virtually every surface around the scene of Ihde's philosophy; quite a bit of Ihde's work can be understood as appropriations and modifications of Heidegger's insights, extracting them from their particular Heideggerian metaphysical context, and developing them into tools for the practical description of human-technology relations. On the other hand, Heidegger serves as a constant point of critique for his allegedly totalizing and dystopian account of technology. Chapter 8, entitled here "The Critique of Heidegger" and selected from his 1993 book *Postphenomenology: Essays in the Postmodern Context*, is one of Ihde's most influential pieces in this line of criticism and includes his parody of Heidegger's classic account of the ancient Greek temple, as Ihde recounts his own experience of viewing nuclear reactor buildings while sailing the Long Island Sound.

As mentioned, another centerpiece notion within the postphenomenological framework of concepts that Ihde has developed is the idea of "multistability." We've seen this above as it had emerged within his account of the variability of human vision and then developed into his conception of our relationships with technology. In chapter 9, entitled "Multistability and Cultural Context" and selected from *Consequences of Phenomenology*, Ihde further explores these kinds of variabilities as technologies develop differently on different parts of the globe. As Ihde reflects on his own preconceptions, we are prompted to think not only on the relativity of technology, but also on Eurocentrism in the philosophy of technology. This chapter also includes his "whole earth measurements" framing perspective, a rhetorical device he has often revisited.

Chapter 10, "The Designer Fallacy," continues these reflections on technological multistability and expands the rhetorical and practical potential of this notion. This chapter, from his 1999 article "Technology and Prognostic Predicaments," is another that I believe to be an underrecognized gem of his corpus, and one that I have come to regularly teach to undergraduate engineering students. This piece contains versions of a number of influential ideas that appear throughout his work, including his warnings against overconfident predictions about technological development, some practical advice for those in design and in the philosophy of technology, and his urging for philosophers to play a stronger role in the development stage of technology (rather than merely engage in after-the-fact technology assessment). It also contains his notion of the "designer fallacy." Making a connection to the intentional fallacy in literary studies, this is his term for the fallacious assumption that a technology will be taken up by users (and will continue to be developed into the future) all along the lines of its designer's expectations.

Part 3: The Phenomenology of Science

Ihde's philosophical reflections on technology continue into an account of the roles of instrumental materiality within scientific practice and epistemology. Through an analysis of the phenomenology of the experience of laboratory instrument usage, he shows the distinct contributions that can be made to both on-the-ground science and the philosophy of science.

Of the canonical phenomenologists, Edmund Husserl was most deeply concerned about issues of epistemology and science. As with his relationship to Heidegger, throughout his career Ihde both builds on Husserl's ideas and uses the figure of Husserl as a constant point of contrast. This fraught relationship culminates in his 2016 book *Husserl's Missing Technologies*. Here in Chapter 11, "The Critique of Husserl," we see another example of Ihde's reevaluation of canonical phenomenology in terms of issues of technology, this time with regard to scientific history and practice. This chapter comes from a 2011 article entitled "Husserl's Galileo Needed a Telescope!," in which Husserl's phenomenological epistemology is taken to task for its alleged lack of engagement with the materiality of scientific practice.

Chapter 12, "Technology Leads Science," selected from his 1983 book *Existential Technics*, continues this line of thinking on the materiality of instrumentation. Where a common understanding may regard technology to merely be a form of applied science, Ihde follows Heidegger in reversing the order. According to Ihde, lines of technological development (such as the continued refinement of, say, microscopes to see smaller and smaller things) provides a major direction of scientific advancement. This idea undermines any account of science as something that moves forward only through theory development and testing. This chapter includes an early version of Ihde's case study of the contrast between seagoing navigation techniques developed within different cultures, one that he would continue to draw on in later works.

Another way that Ihde has followed out this line of thinking is through the development of the notion of "epistemology engines," that is, technological metaphors that guide scientific, philosophical, and design thinking. While the computer may play such a role today (serving as a guiding metaphor for everything from human thought to the genetic code), Ihde conducts a case study of the history of what he argues is one of the most central epistemology engines in Western history: the camera obscura. Ihde shows in detail how this device guided thought about human perception, the mind, scientific methodology, and the specifics of scientific experimentation. Chapter 13, "Epistemology Engines and the Camera Obscura," comes from one example of Ihde's continuing studies of this device, a 2008 book chapter entitled "Art Precedes Science: Or Did the *Camera Obscura* Invent Modern Science?"

Chapter 14, "The Phenomenology of Scientific Imaging," presents some of Ihde's most detailed and influential ideas regarding the phenomenology of scientific instrumentation: his analysis of laboratory imaging. In ideas taken up by a generation of philosophers and anthropologists studying scientific practice, Ihde has made the topic of imaging one of his career's central points of fascination.[3] This chapter comes out of two from his 1998 book *Expanding Hermeneutics: Visualism in Science*, the final third of which he has referred to as a "minimonograph" on this theme. This work involves a detailed phenomenology of the hermeneutic and gestalt experience of scientists' perception of meaning within images. And Ihde follows out the implications for both practical issues of scientific praxis and philosophical issues of scientific realism.

The project of putting together this volume has been something of a labor of love. More than just the expected tasks of selecting pieces, reformatting, and chasing down permissions, this has been a job of wrestling with word processing. The original computer files for these pieces are simply unavailable. So the task of getting these texts into this volume often involved physically scanning them from the printed books, and then translating the digital text from those scans, a process that inevitably introduced all manner of errors and gobbledygook in need of fixing—slowly, line by line—original texts in-hand. (Surely a more tech savvy person could have developed a better process. But this is what I had available.) I am happy to do it. Ihde's works, as well as his mentorship and our close friendship, have meant so much to me and my life as a scholar. I hope this volume can help others to find joy and excitement in their own work.

References

Aagaard, Jesper, Jan Kyrre Berg Friis, Jessica Sorenson, Oliver Tafdrup, and Cathrine Hasse, eds. 2018. *Postphenomenological Methodologies: New Ways in Mediating Techno-Human Relationships.* Lanham, MD: Lexington Books.

Botin, Lars, Bas de Boer, and Tom Børsen, eds. 2020. "Technology in between the Individual and the Political: Postphenomenology and Critical Constructivism." Special issue, *Techné* 24, no. 1/2.

Botin, Lars, and Inger Berling Hyams, eds. 2021. *Postphenomenology and Architecture: Human Technology Relations in the Built Environment.* Lanham, MD: Lexington Books.

de Boer, Bas. 2021. *How Scientific Instruments Speak: Postphenomenology and Technological Mediations in Neuroscientific Practice.* Lanham, MD: Lexington Books.

Fried, Samantha J., and Robert Rosenberger, eds. 2021. *Postphenomenology and Imaging: How to Read Technology.* Lanham, MD: Lexington Books.

Friis, Jan Kyrre Berg, and Robert P. Crease, eds. 2015. *Technoscience and Postphenomenology: The Manhattan Papers.* Lanham, MD: Lexington Books.

Goeminne, Gert, and Erik Paredis, eds. 2011. "Opening Up the In-Between: Interdisciplinary Reflections on Science, Technology and Social Change." Special issue, *Foundations of Science* 16, no. 2–3.

Hasse, Cathrine. 2020. *Posthumanist Learning: What Robots and Cyborgs Teach Us about Being Ultra-social.* London: Routledge.

Ihde, Don. 1971. *Hermeneutic Phenomenology: The Philosophy of Paul Ricoeur.* Evanston, IL: Northwestern University Press.

———. 1973. *Sense and Significance.* New York: Humanities Press.

———, ed. 1974. *The Conflict of Interpretations: Paul Ricoeur.* Evanston, IL: Northwestern University Press.

———. 1976. *Listening and Voice: A Phenomenology of Sound.* Athens: Ohio University Press.

———. 1977. *Experimental Phenomenology.* New York: Putnam. Reprinted 1986, State University of New York Press.

———. 1979. *Technics and Praxis.* New York: Reidel.

———. 1983. *Existential Technics.* Albany: State University of New York Press.

———. 1986a. *Consequences of Phenomenology.* Albany: State University of New York Press.

———. 1986b. *On Non-foundational Phenomenology.* Edited by Seth Chaiklin. Fenomenografiska notiser 3. Gothenburg: Institutionen för pedagogik, Göteborgs universitet.

———. 1990. *Technology and the Lifeworld: From Garden to Earth.* Bloomington: Indiana University Press.

———. 1991. *Instrumental Realism: The Interface between Philosophy of Science and Philosophy of Technology.* Bloomington: Indiana University Press.

———. 1993a. *Philosophy of Technology: An Introduction.* New York: Paragon House.

———. 1993b. *Postphenomenology: Essays in the Postmodern Context.* Evanston, IL: Northwestern University Press.

———. 1998. *Expanding Hermeneutics: Visualism in Science.* Evanston, IL: Northwestern University Press.

———. 1999. "Technology and Prognostic Predicaments." *AI and Society* 13:45–51.

———. 2002. *Bodies in Technology.* Minneapolis: University of Minnesota Press.

———. 2007. *Listening and Voice: Phenomenologies of Sound.* 2nd ed. Albany: State University of New York Press.

———. 2008a. "Art Precedes Science: Or Did the Camera Obscura Invent Modern Science?" In *Instruments in Art and Science: On the Architectonics of Cultural Boundaries in the Seventeenth Century,* edited by Helmar Schramm, Ludger Schwarte, and Jan Lazardzig, 383–93. Berlin: de Gruyter.

———. 2008b. *Ironic Technics.* New York: Automatic Press/VIP.

———. 2008c. *Let Things Speak* [in Chinese]. Translated by H. Lianqing. Peking: Peking University Press.

———, ed. 2008d. "Postphenomenological Research." Special issue, *Human Studies* 31, no. 1.

———. 2009. *Postphenomenology and Technoscience: The Peking University Lectures.* Albany: State University of New York Press.

———. 2010a. *Embodied Technics.* New York: Automatic Press/VIP.

———. 2010b. *Heidegger's Technologies: Postphenomenological Perspectives.* New York: Fordham University Press.

———. 2011. "Husserl's Galileo Needed a Telescope!" *Philosophy and Technology* 24:69–82.

———. 2012. *Experimental Phenomenology: Multistabilities.* 2nd ed. Albany: State University of New York Press.

———. 2015. *Acoustic Technics*. Lanham, MD: Lexington Books.

———. 2016. *Husserl's Missing Technologies*. New York: Fordham University Press.

———. 2019. *Medical Technics*. Minneapolis: Minnesota University Press.

———. 2021. *Material Hermeneutics: Reversing the Linguistic Turn*. London: Routledge.

Ihde, Don, and Lambros Malafouris, eds. 2019. "Homo Faber Revisited: Postphenomenology and Material Engagement Theory." Special issue, *Philosophy and Technology* 32, no. 2.

Ihde, Don, and Evan Selinger, eds. 2003. *Chasing Technoscience: Matrix for Materiality*. Bloomington: Indiana University Press.

Ihde, Don, and Hugh J. Silverman, eds. 1985. *Descriptions*. Selected Studies in Phenomenology and Existential Philosophy 11. Albany: State University of New York Press.

Ihde, Don, and Richard M. Zaner, eds. 1975. *Dialogues in Phenomenology*. Selected Studies in Phenomenology and Existential Philosophy 5. The Hague: Martinus Nijhoff.

———, eds. 1977. *Interdisciplinary Phenomenology*. Selected Studies in Phenomenology and Existential Philosophy 6. The Hague: Martinus Nijhoff.

Irwin, Stacey O'Neal. 2017. *Digital Media: Human-Technology Connection*. Lanham, MD: Lexington Books.

Lemmens, Pieter, and Yoni Van Den Eede, eds. 2022. "Rethinking Technology in the Anthropocene." Special issue, *Foundations of Science* 27, no. 1.

Miller, Glen, and Ashley Shew, eds. 2020. *Reimagining Philosophy and Technology, Reinventing Ihde*. Cham, Switzerland: Springer.

Rosenberger, Robert, ed. 2016. "Ihde and Husserl: A Symposium on Ihde's *Husserl's Missing Technologies*." Special issue, *Techné* 20, no. 2.

———. 2017. *Callous Objects: Designs against the Homeless*. Minneapolis: University of Minnesota Press.

Rosenberger, Robert, and Peter-Paul Verbeek, eds. 2015. *Postphenomenological Investigations: Essays on Human-Technology Relations*. Lanham, MD: Lexington Books.

Selinger, Evan, ed. 2006. *Postphenomenology: A Critical Companion to Ihde*. Albany: State University of New York Press.

———, ed. 2008. "Postphenomenology: Historical and Contemporary Horizons." Special issue, *Techné* 12, no. 2.

Silverman, Hugh J., and Don Ihde, eds. 1985. *Hermeneutics and Deconstruction*. Selected Studies in Phenomenology and Existential Philosophy 10. Albany: State University of New York Press.

Van Den Eede, Yone, Stacey O'Neal Irwin, and Galit P. Wellner, eds. 2017. *Postphenomenology and Media: Essays on Human-Media-World Relations*. Lanham, MD: Lexington Books.

Verbeek, Peter-Paul. 2011. *Moralizing Technology: Understanding and Designing the Morality of Things*. Chicago: University of Chicago Press.

Wellner, Galit P. 2015. *A Postphenomenological Inquiry of Cell Phones.* Lanham, MD: Lexington Books.

Zaner, Richard M., and Don Ihde, eds. 1973. *Phenomenology and Existentialism.* New York: Capricorn Books.

Part 1

From Phenomenology to Postphenomenology

Chapter 1

Auditory Phenomenology

The Auditory Dimension

What is it to listen *phenomenologically*? It is more than an intense and concentrated attention to sound and listening, it is also to be aware in the process of the pervasiveness of certain "beliefs" that intrude into my attempt to listen "to the things themselves." Thus the first listenings inevitably are not yet fully existentialized but occur in the midst of preliminary approximations.

Listening begins with the ordinary, by proximately working its way into what is as yet unheard. In the process the gradual deconstruction of those beliefs that must be surpassed occurs. We suppose that there are significant contrasts between sight and sound; thus in the very midst of the implicit sensory atomism held in common belief we approximate abstractly what the differences might be between the dimensions of sight and of sound.[1] We "pair" these two dimensions comparatively. First we engage in a hypothetical and abstract mapping that could occur for ordinary experience with its inherent beliefs.

Supposing now two "distinct" dimensions within experience that are to be "paired," I attend to what is seen and heard to learn in what way these dimensions differ and compare, in what ways they diverge in their respective "shapes," and in what ways they "overlap."

I turn back, this time imaginatively, to my visual and auditory experience and practice a kind of free association on approximate visual and auditory possibilities, possibilities not yet intensely examined, which float in a kind of playful reverie.

Before me lies a box of paper clips. I fix them in the center of my vision. Their shape, shininess, and immobility are clear and distinct. But as soon as I pair their appearance with the question of an auditory aspect I note that they are also *mute*. I speculatively reflect on the history of philosophy with recollections of pages and pages devoted to the discussion of "material objects" with their various qualities and on the "world" of tables, desks, and chairs that inhabit so many philosophers' attentions: *the realm of mute objects*. Are these then the implicit standard of a visualist metaphysics? For in relation to stable, mute objects present to the center of clear and distinct vision, the role of *predication* seems easy and most evident. The qualities adhere easily to these material objects.

A fly suddenly lands on the wall next to the desk where the paper clips lie and begins to crawl up that wall. My attention is distracted and I swat at him. He quickly, almost too quickly for the eye, escapes and flies to I know not where. Here is a moving, active being on the face of the visual "world." With the moving, active appearance of the fly a second level or grouping of objects displays itself. This being, which is seen, is active and is characterized by motion. Movement belongs to the verb. *He walks, he flies, he escapes.* These are not quite correctly properties but activities. Who are the "metaphysicians" of the fly? I recall speculatively those traditions of "process" and movement that would question the dominance of the stable, mute object, and see in motion a picture of the world. The verb is affirmed over the predicate.

But the metaphysicians of muteness may reply by first noting that the moving being appears against the background of the immobile, that the fly is an appearance that is discontinuous, that motion is an occasional "addition" to the stratum of the immobile. The fly's flight is etched against stability, and the arrow of Zeno, if it may speed its way at all, must do so against the ultimate foundation of the stable background. Even motion may be "reduced" to predication as time is atomized.

But what of sound? The mute object stands "beyond" the horizon of sound. Silence is the horizon of sound, yet the mute object is silently *present*. Silence seems revealed at first through a visual category. But with the fly and the introduction of motion there is the presentation of a buzzing, and Zeno's arrow whizzes in spite of the paradox. Of both animate and inanimate beings, motion and sound, when paired, belong together. "Visualistically" sound "overlaps" with moving beings.

With sound a certain liveliness also makes its richer appearance. I walk into the Cathedral of Notre Dame in Paris for the first time. Its emptiness and high arching dark interior are awesome, but it bespeaks a certain monumentality. It is a ghostly reminder of a civilization long past, its muted walls echoing only the shuffle of countless tourist feet. Later I return, and a high mass is being sung: suddenly the mute walls echo and reecho and the singing fills the cathedral. Its soul has momentarily returned, and the mute testimony of the past has once again returned to live in the moment of the ritual. Here the paired "regions" of sight and sound "synthesize" in dramatic richness.

But with the "overlapping" of sight and sound there remains the "excess" of sight over sound in the realm of the mute object. Is there a comparable area where listening "exceeds" seeing, an area beyond the "overlapping" just noted where sight may not enter, and which, like silence to sound, offers a clue to the horizon of vision?

I walk along a dark country path, barely able to make out the vague outlines of the way. Groping now, I am keenly aware of every sound. Suddenly I hear the screech of an owl, seemingly amplified by the darkness, and for a moment a shock traverses my body. But I cannot see the bird as it stalks its nocturnal prey. I become more aware of sound in the dark, and it makes its presence more dramatic when I cannot see.

But night is not the horizon of sight, nor Dionysius the limit of Apollo. I stand alone on a hilltop in the light of day, surveying the landscape below in a windstorm. I hear its howling and feel its chill but I cannot see its contorted writhing though it surrounds me with its invisible presence. No matter how hard I look, I cannot see the wind, *the invisible is the horizon of sight*. An inquiry into the auditory is also an inquiry into the invisible. Listening makes the invisible *present* in a way similar to the presence of the mute in vision.

What metaphysics belong to listening, to the invisible? Is it also that of Heraclitus, the first to raise a preference for vision, but who also says, "Listening not to me but to the Logos, it is wise to acknowledge that all things are one" (Wheelwright 1966, 79). Is such a philosophy possible beyond the realm of mute objects? Or can such a philosophy find a way to give voice even to muteness? The invisibility of the wind is indicative. What is the wind? It belongs, with motion, to the realm of verb. The wind is "seen" in its *effects*, less than a verb, its visible being is what it has done in passing by.

Is anything revealed through such a playful association? At a first approximation it seems that it is possible to map two "regions" which do not coincide, but which in comparison may be discerned to have differing boundaries and horizons. In the "region" of sight there is a visual field which may be characterized now as "surrounded" by its open horizon which limits vision, and which remains "unseen." Such a field can be diagrammed (fig. 1.1).

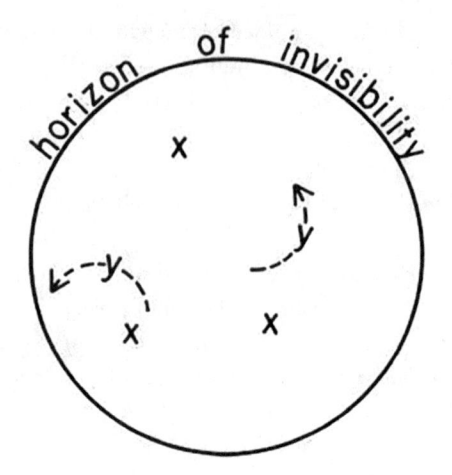

Figure 1.1. Horizon of Invisibility. Drawing created by Don Ihde.

Here, where the enclosed circle is the present visual field, within this presence there will be a vast totality of entities that can be experienced. And although these entities display themselves with great complexity, within the abstraction of the approximation we note only that some are stable (x) and usually mute in ordinary experience, and that some (-y--) move, often "accompanied" by sounds. Beyond the actually seen field of presence lies a horizon designated now as a horizon of invisibility.

A similar diagram can be offered for a "region" of sound presences (fig. 1.2).

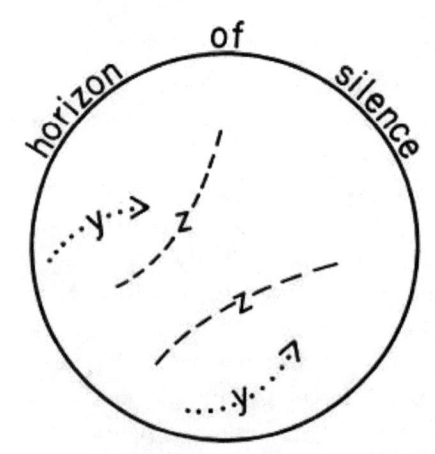

Figure 1.2. Horizon of Silence. Drawing created by Don Ihde.

Although once we move beyond this approximation, the "shape" of the auditory field will need to be qualified. Within the limits of the first approximation we note that the auditory field contains a series of auditory presences which do not, however, perfectly overlap those of the visual field. There are sounds that "accompany" moving objects or beings (-y-), but there are some for which no visible presence may be found (--z--). Insofar as all sounds are also "events," all the sounds are, within the first approximation, likely to be considered as "moving." Again, there is also a horizon, characterized by the pairing as a horizon of silence that "surrounds" the field of auditory presence.

It is also possible to relate, within the first approximation, the two "regions" and discern that there are some overlapping and some nonoverlapping features of each "region." Such a "difference" may be diagrammed (fig. 1.3).

In this diagram of the overlapping and nonoverlapping "regions" of sight and sound we note that what may be taken as horizonal (or absent) for one "region" is taken as a presence for the other.

Thus while the area of mute objects (x) seems to be closed to the auditory experience as these objects lie in silence, so within auditory experience the invisible sounds (--z--) are present to the ear but absent to the eye. There are also some presences that are "synthesized" (-y--) or present to both "senses" or "regions."

This pairing when returned to the reverie concerning the associated "metaphysics" of the "senses" once more reveals a way in which the -traditions

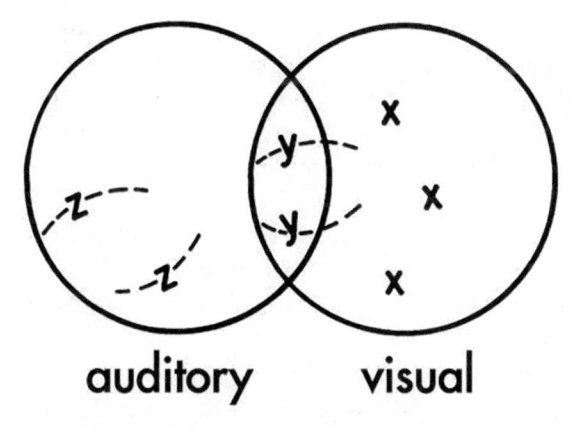

Figure 1.3. Auditory-Visual Overlap. Drawing created by Don Ihde.

of dominant visualism show themselves. If we suppose that any metaphysics of worth must be one that is at least comprehensive, then a total visualist metaphysics must find a way to account for and to include in its description of the world all those invisible events that at this level seem to lie beyond the reach of the visible horizon, but are nevertheless present within experience.

This may be done in several ways. First, one can create some hermeneutic device which, continuing the approximation of the "regions," functionally makes the invisible visible. This implies some "translation" of one "region" into the terms of the favored "region." Such is one secret of the applied metaphysics often found in the sciences of sound. Physically, sound is considered a wave phenomenon. Its wave characteristics are then "translated" into various visual forms through instruments, which are the extended embodiments of the scientific enterprise. Voice patterns are "translated" into visual patterns on oscillographs; sound reverberations are mapped with Moire patterns; even echolocation in its practical applications is made a matter of seeing what is on the radar screen: the making or "translating" of the invisible into the visible is a standard route for understanding a physics of sound.

In the case of the sciences of sound this translation allows sound to be measured, and measurement is predominantly a matter of spatializing qualities into visible quantities. But in ordinary experience there is often thought to be a similar role for sound. Sounds are frequently thought of as anticipatory clues for ultimate visual fulfillments. The most ordinary of such occurrences are noted in locating unseen entities.

The bird-watcher in the woods often first hears his bird, then he seeks it and fixes it in the sight of his binoculars. The person hanging a picture knows where to look for the dropped tack from the sound it made as it rolled under the piano. And although not all noises yield a visual presence—for example, the extreme case of radio astronomy may yield the presence of an unsuspected "dark" star that may never be seen—the familiar movement from sound to sight may be discerned.

The movement from that which is heard (and unseen) to that which is seen raises the question of its counterpart. Does each event of the visible world offer the occasion, even ultimately from a sounding presence of mute objects, for silence to have a voice? Do all things, when *fully* experienced, also sound forth?

In ordinary experience this direction is also taken. The bird-watcher may be an appreciative bird listener. He awaits quietly in the hopes that the winter wren will sing his long and complicated "Mozart" song. But only in more recent times has this countermovement become conspicuous. The amplified listening which now reveals the noise of lowly ant societies gives voice to the previously silent. Physically even molecules sound, and the human ear comes to a threshold of hearing almost to the point of hearing what would be incessant noise (von Bekesy 1967, 9).

The Shapes of Sound

The approximation that opened a difference between sight and sound ended in a questioning of the import of that difference. If a movement is possible that gives visibility to the unseen, and a countermovement that gives voice to the mute is possible, a closer listening to the auditory dimension itself is called for. The time has come when listening must begin to be reflective. I begin to take note of my listening, and I first notice a certain incessant field of sound which strikes me as a constant "flux" marked by an obvious and dramatic "temporality."

I begin to catalog my auditory experience within a given moment of time, and I note that within only a few moments a series of sound-events have occurred. There is the sound of the vacuum cleaner on the floor below; just then there was the pounding of the construction worker next door; the rustle of leaves is heard momentarily; and, if I am more attentive to less obvious sounds, there is the buzz of the fluorescent light and the hum of the heating system. But I also conclude, perhaps too easily and too quickly, that the auditory world is one of "flux" and that it is primarily *temporal.*

I close my eyes and note that one sound follows another, that a single sound "exists" for a moment and "passes away," and that there is an "inconstancy" to this "region" in which the surging of time is dramatically present. This intimacy of temporality with the auditory experience forms a central tradition concerning sound and may be found recorded by philosophers as diverse in points of view as Søren Kierkegaard, Edmund Husserl, and P. F. Strawson. What I "discover" first is already known and sedimented as knowledge.

In meditating on music and language as sensuous media Kierkegaard writes, "The most abstract idea conceivable is sensuous genius. But in what medium is this idea expressible? Solely in music . . . it is an energy, a storm, impatience, passion, and so on, in all their lyrical quality" (Kierkegaard 1959, 1:55). But also noting the auditory dimension of language he states, "Language addresses itself to the ear. No other medium does this. The ear is the most spiritually determined of the senses . . . aside from language, music is the only medium that addresses itself to the ear" (Kierkegaard 1959, 1:66).

Language and music, auditory phenomena, are understood by Kierkegaard to be dominantly temporal in their actual form. "Language has time as its element; all other media have space as their element. Music is the only other one that takes place in time" (Kierkegaard 1959, 1:67). This *positive* relation of sound to time is what contextually appears as "first" in a reflective listening. It is also maintained within phenomenology in the use of auditory material in Husserl's *Phenomenology of Internal Time Consciousness.* Not only are his usual visual examples often and even dominantly replaced in these lectures by auditory ones, but even the use of metaphorical and descriptive language begins to take on an auditory tone. "The bird changes its place; it flies. In every situation the echo of earlier appearances clings to it (i.e., to its appearance). Every phase of this echo, however, fades while the bird flies farther on. Thus a series of *'reverberations'* pertains to every subsequent phase, and we do not have a simple series of successive phases" (Husserl 1964, 149; italics mine). With Kierkegaard, Husserl takes note of the overwhelming intimacy of sound and time.

However, where this is "traditional" concerning sound, and where this strong tie cannot be overlooked in any analysis of auditory experience, there is often either implicitly or explicitly a negative claim that listening is either therefore "weak" spatially or, most extremely, that *sound lacks spatiality entirely.* This negative claim is most blatant in precisely that tradition that most clearly "atomizes" the senses and reduces them to their lowest forms: the empiricist tradition. This view, rapidly losing ground in the biological and physical sciences, sometimes affirms that "a spatial order is connate only for the optical, tactile, and kinesthetic spheres; while for the other senses, mere complexes of feelings with spatial features are admitted" (Straus 1966, 4).

Such a view is explicit in Strawson's *Individuals.* Strawson, clearly defending a "metaphysics of objects" in an Aristotelian vein that continues the visualism of vision and objectification, considers a "No-Space" world that he finds a conceptual possibility of a reduced world of "pure" sound. "The fact is that where sense experience is not only auditory in character, but also at least tactual and kinesthetic as well—we can sometimes as *sign* spatial predicates on the strength of hearing alone. But from this fact it does not follow that where this experience is supposed to be exclusively auditory in character, there would be any place for spatial concepts at all. *I think it is obvious* that there would be no such place" (Strawson 1971, 65; italics mine).

What is "obvious" is that a tradition is here being taken for granted with disregard for the contemporary discoveries of very complex spatial attributes to auditory experience. Directionality and location, particularly advanced in such animals as porpoises and bats but not lacking in humans, have shown the degree to which echolocation is a very precise spatial sense. The bat's ability to "focus" a "ray" of sound such that it may discern the difference between a twig and the moth it is after is now well known. But such auditory abilities have long been encased in precisely the tradition that denies spatiality to listening and for decades and even centuries prevented the scientist from believing that it was indeed a capacity for sound and listening.

Although experiments with bats as early as those of Lazzaro Spallanzani in 1799 led him to ask "whether their ears rather than their eyes serve to guide them in flight." The already established prejudices of the ancients caused even Spallanzani to doubt his findings. Even the suggestion that hearing could detect and localize objects "in space" was vigorously attacked by eminent figures such as Georges Cuvier and George Montagu. It was not until 1912 that the suggestion that hearing was "spatializable" reopened the question that has led to contemporary knowledge concerning echolocation in a whole series of animals, and that today may lead to the development of amplification devices by which the blind may extend their often already acute hearing.

It is precisely because of the very "obviousness" *not* of experience, but of the traditions concerning experience that there is reason to postpone what is "first" in the turn to the auditory dimension. Without denying the intimacy of sound and time and without denying the richness of the auditory in relation to temporality, a strategy that begins in approximations is one that must move with extreme care so as not to overlook or fail to hear what also may be shown in the seemingly weaker capacities of auditory experience. Thus as the move into phenomenology proper is made, it is with the *spatiality* of sound that description may begin. Within a spirit of gradual approximations the "weakest" possibilities of sound are to be explored before the "strongest" possibilities.

However, there are several initial qualifications that must be held in mind in beginning phenomenological description in this way. First, the movement from the more abstract approximation that began in the midst of sensory atomism is one that not only increasingly accelerates away from that division of the senses, but one that begins in making thematic what will be called here the first existential level of experience, the level of "greatest naïveté." For despite the extreme technicality of Husserl's discussion of identity from *Logical Investigations*, the outcome is one that reaffirms the primacy of the *thing* in naive or first existential experience. It is to *things* that we attend in naive and ordinary experience once we have set aside our layers of beliefs regarding how those things "should" present themselves.

The same applies to auditory experience. Sounds are "first" experienced as sounds of things. That was the sound *of* the jackhammer with all its irritating intrusion. There, it's Eric calling Leslie now. That was definitely a truck that went by rather than a car. This ease that we take for granted and by which we "identify" things by sound is part of our ongoing ordinary experience. This common ability of listening contains within it an extraordinary richness of distinction and the capacity to discern minute differences of auditory texture, and by it we know to what and often to where it is that our listening refers.

Often we find extraordinary examples of these capacities in the *musical ear*. Beethoven, for example, had such a rich and extraordinary auditory ability, both perceptual and imaginative, that he could compose and imaginatively hear a whole symphony in his head and specifically discouraged anyone from using the piano to demonstrate passages, because the piano was much poorer than the whole symphony in his head. But this musical, perceptual memory, though not equally acute, is not rare among accomplished musicians.

Such musical feats are also potentially misleading, because there is also a tradition, echoed above by Kierkegaard, that music is "abstract." Even phenomenologists have been misled to take the musical experience as one that is disembodied and "separated from its source" as a kind of "pure" auditory experience (Straus 1966, 7).

In daily concerns such abstract listening is at least unusual, yet its feats of discernment are highly discriminating. On walking along a village street in Llangefni, Wales, my son pointed out a thrush busily banging a snail against the sidewalk. This act soon successfully produced a tasty meal even without benefit of garlic and butter. Several weeks later I was awakened in our house in London to the early morning unmistakable cracking of the snail shell coming through the curtained window. I drew the curtains to show my wife this occurrence, which was new to her, but the "identification" had been quite "obvious" to me by the single sound of the cracking snail shell.

Such identifications and discriminations of minute auditory differences are not as yet "spatial." But having made a turn of attention to the first naive existential level of experience where sounds are the sounds of things, the spatial aspects of that experience may begin to show themselves. In searching out the spatiality of sound the cautions previously noted take specific form here. First, auditory spatiality must be allowed to "present itself" as it "appears" within this level of experience. Negatively, a predefinition of spatiality such that it is prejudged "visualistically" must be suspended.

Second, affirming the phenomenological sense of the global character of primal experience, it is necessary to replace the division of the senses with the notion of a *relative focus* on a dimension of global experience such that it is noted only against the omnipresence of the globality. Thus a "pure" experience is eliminated and made impossible. Primitively things are always already found "synthesized" in naive existential experience. The move to a focus fringe interpretation of global experience thus safeguards the tendency toward disembodiment that tempts all "Cartesian" types of philosophy, and mixes, in spite of itself, perceptual and emptily suppositional terms.

Third, as a first phenomenological approximation in contrast to the approximation in the midst of sensory atomism, it should be noted that even the division of space and time are not, strictly speaking, primitive experiential significations. Existentially there is a concrete space-time that is also a signification of naive experience in its thematized appearance.

With this second approximation, the entry into the "weakness" of the auditory dimension, phenomenological description proper begins. The provisional character of the sounds of things in ordinary experience should not be considered a final but a first level of the phenomenological experiential analysis.

Shapes, Surfaces, and Interiors

At the experiential level where sounds are heard as the sounds of things it is ordinarily possible to distinguish certain *shape-aspects* of those things. The following variations begin in what for human hearing is admittedly one of the weakest existential possibilities of listening. I do not claim that every sound gives a shape-aspect (but neither does every sighting give a shape-aspect in the ordinary sense). At first such an observation seems outrageous: *we hear shapes.*

The shape-aspects that are heard, however, must be strictly located in terms of their auditorily proper presentation and not predetermined or prelimited by an already "visualist" notion of shape. The shape-aspects that are heard are "weaker" in their spatial sense than the full outline shape of a thing that is ordinarily given all at once to vision. But a "weakness" is not necessarily a total absence, for in this "weakness" there remains an important, if primitive, spatiality for hearing.

Children sometimes play an auditory game. Someone puts an object in a box and then shakes and rolls the box, asking the child what is inside. If, more specifically, the question is directed toward shapes, the observer soon finds that it takes little time to identify simple shapes and often the object by its sound. For example, if one of the objects is a marble and the other a die (of a pair of dice), and the box is rolled, the identification is virtually immediate. The difference of shape has been *heard,* and the shape-aspect has been auditorily discriminated.

But the flood of likely objections to such an observation, however experientially concrete such examples are, threatens to overwhelm the listener. For in spite of the hermeneutic rules of *epoché* that attempt to put out of play both "sensory atomism" and its preferred "visualism," it threatens to return at each stage of analysis. It is precisely the recalcitrance of such beliefs that makes the act of auditory discernment "difficult to believe" in spite of one's ears.

The point here is not to enter into an interminable and difficult argument but to let the things show or sound themselves. For involved in the "weakness" of auditory spatiality there are a number of factors that allow that "weakness" to be missed if one is not careful in listening. What is amazing, however, is what appears spontaneously in the simple variation. The very first time I played this game with my son I had placed a ballpoint pen in a box without his seeing it and rolled it back and forth. I asked him what shape it was. His answer was, "It's like a fifty-pence shape, you know, on its sides, only it's longer." A fifty-pence coin has seven sides, the ballpoint pen had six, and it was, in his parlance, "longer."

The shape-aspect is not the only thing that is given in the richness of simple auditory presentations. If the game is allowed to continue so that one learns to hear things in an analogue to the heightened hearing of the blind man's more precise listening to the world, a quickly growing sophistication occurs. A ballpoint pen gives a quite different auditory presentation with its plastic click from that of a wooden rod. A rubber ball is as auditorily distinct from a billiard ball as it is visually distinct. The very texture and composition as well as the shape-aspect is presented in the complex richness of the event.

It is often this learning itself that offers itself as suspicious to the "sensory atomist" whose notion of a built-up or constructed knowledge also infects his understanding of learning. Phenomenologically there is a great distinction between *constructing* something and its *constitution*. In constitution the learning that occurs is a learning that becomes aware of what there is to be seen or heard. There is the usual inversion called for in *epoché* here. As Merleau-Ponty remarked, "*Learning is In der Welt Sein, and not at all that In der Welt Sein is learning*" (Merleau-Ponty 1968, 212). *This* difference may be illustrated by two vastly different ways in which perceptual experience is employed in the empirical sciences.

In some psychology many of the experiments are deliberately designed to first disrupt all previous "learning" by radically altering its context. To view a white sheet of paper under blue lighting through a darkened tube that cuts off the normal context and field significance of the experience is to radically alter ordinary experience. But the learning which is tacit in ordinary experience is then further cut off by allowing the experience to continue for only an atom of time, thus preventing any adjustment. In this way the experiment is set up so that it often cannot help but circularly reinforce the "abstraction" of the "sensory atomist's" view of perception that begins with the "abstraction" of "sense data" or similar "stimuli." The experiment constructs the condition for the preformed conclusion and interprets what it finds as *a primitive* of experience.

Yet the always-present learning through which perceptions are incarnate functions here as well; only in this case it operates tacitly in the situation of the observer, the psychologist. Were she to be replaced by another observer as experimentally "naive" as her subject, in all likelihood there would be little purpose to or knowledge resulting from the experiment. The observer would still have to enter the scene even if now her taken-for-granted judgments are removed one step further.

There is a sense in which the role of constitution proposed by the phenomenologist is implicitly recognized in the natural sciences. An ornithologist friend once described to me the pains he had to go through to get his first-year students to even produce a recognizable description of bird behavior. He would lapse into laughs when a report returned stating, "The bird *sat* on the fence, then it hopped and sat again." For in his parlance a bird not only does not sit, it perches; but the student in this case had not yet learned to see. At first the learner does not recognize the differences between the various species of warblers, which are often confusing anyway, but after long and careful learning he then wonders why he could not at first recognize what is now so obvious.

But the learning does not *construct* what is to be seen, it *constitutes* it in terms of its meaning. What is to be seen is *there*, and anyone entering this region of knowledge may see the distinctive marks that differentiate one warbler from the other. Once the distinctions have been learned, the previous lack of awareness and lack of discrimination is seen not as a fault of the "object" but of the inadequacy of our own prior observation.

This problem is partially due to our frequent failure to discern the space-aspects of auditory experience. We have not learned to listen for shapes. The whole of our interpretation in its traditional form runs from it, and only in the dire situation of being forced to listen for shapes, such as the advent of blindness, do ordinary men attend specifically to the shape aspects of sound.

But even here there is a complication that arises from the global or plenary quality of primary experience. For the blind or deaf person experiences his "world" as a unity and his experience as a plenum. His sense of lack is conveyed by the transcendence of language, and he even becomes quite adept at "verbalisms," the ability to define things through words, although he may not recognize them when they are presented to him. One blind person describes this sense. "Those who see are related to me through some unknown sense which completely envelops me from a distance, follows me, goes through me, and, from the time I get up to the time I go to bed, holds me in some way in subjection to it" (Heaton 1968, 42).

It is here that the "sensory atomist" finds so much "evidence" for his constructionist view of the world. It is well known that many, indeed most, persons who are blind cannot visually recognize certain objects presented to them until they feel these objects. But there is another possible interpretation of such "evidence." It is not that the object is built up, but that the learning that goes on in all experience must go on here, too. The radically new experience of seeing, when a blind person gains sight through a medical procedure, is revealing. His *first* sight, when reported, often turns out to be precisely "like" those first impressions reported in the first turn to reflective *listening*. He is impressed by what we might call the *flux* and *flow*, the implicit temporality of the new dimension to his experience. J. M. Heaton reports that when the blind are given sight, "at first colours are not localized in space and are seen in much the same way as we smell odours" (Heaton 1968, 42). Odors, sounds, tastes, on *first* note, appear not as fixed, but as a flux and flow. The first look is a stage of experience, not something that belongs isolated within one "sense."

This learning is often painful. For the patient it is not a mere addition to his experience but a transformation of the whole previous shape of the plenum of his experience: "The chief difficulty experienced by these patients is due to the general reorganization of their existence that is required, for the whole structure of their world is altered and its centre is displaced from touch to vision; and not only perception but language and behavior also have to be reoriented" (Heaton 1968, 43).

It is not, however, that on being given sight spatiality is first discovered. It is *reconstituted*. A subtle example of this was given to me by a student trained in phenomenology who had been blind, but who, through treatment, gained limited sight. She noted that one quite detectable difference in her lived spatial organization when given sight was a gradual displacement of a previously more omnidirectional orientation and spatial awareness to a much more focused *forward* orientation. Although she noted that even when blind there was a slight "preference" for a forward directed awareness, this became much more pronounced with the gaining of vision (Haughawout 1969, 4). Again, as will become more apparent as the spatial significations of the auditory dimension become more pronounced, the relative omnidirectionality of awareness and orientation is "closer" to the space-sense of sound than that of vision.

In a gradual clarification of the distinctive spatial sense of auditory experience, the first discrimination of shape-aspects heard in such spontaneous experiences as that of the game of placing an object in a box becomes more precise when attention is paid not only to the presence of the spatial aspect, but to how it is given in perception. Reverting to the pairing of sight and sound, this factor becomes easier to locate.

I turn to my visual and auditory experiences. I note now that in both dimensions there is a multiplicity of phenomena, but I also note that these do not always overlap. I see before me the picture of the sailboat, the note concerning last night's sherry party, a postcard from Japan. But I hear the cement mixer, the bird song, and the traffic in the street.

Next, I note that it seems at first that every stable thing before me visually presents a spatial signification which is, moreover, given-all-at-once. Each object has at least an *outline shape*, and this shape in the objects mentioned is discerned immediately. But of the sounds I do not seem to get shapes, certainly not outline shapes and certainly not all-at-once.

In comparing this nonoverlapping of shape in sight and sound in terms of the question of how shape-aspects are given, I soon find that the question of time is involved as well. The all-at-onceness to the outline shape before me is a matter of *temporal instantaneousness* or of *simultaneity*. But when I return to those experiences which give me shape-aspects I find that the one given is not a matter of instantaneousness but of a *sequential* or *durational* presentation. If the ball is dropped and does not bounce, I may not get more than a "contact point" as a vague and extremely "narrow" signification. But if the ball is rolled for several instants, if the rolling endures through a time span which is quite short, I get a sense of its shape as an *edge-shape*. This shape is presented not in terms of temporal instantaneousness but in terms of temporal duration. In both cases there is a need for some "time," as even visually the object presented in too small an atom of time remains equally spatially indiscernible. But there is a difference of need here in which the temporal duration for the discrimination of an edge-shape by sound must be relatively greater. Here again a clue seems to emerge as to why tradition has maintained the asymmetries of "spatial poverty" for sound and "temporal richness" for sound in comparison to the "spatial richness" and "temporal poverty" for sight.

But this comparative variation bespeaks only one, albeit important, variation in relation to spatial significations, and with it the sedimentation of the dominance of the mute object for spatial significations remains. Further variations, however, tend to diminish the asymmetries to a degree. If I return to the pairing of sight and sound and introduce the (rapidly) moving thing into experience, a difference occurs. The arrow, the drop of water, the stone that appear before me falling or flying at certain speeds do not show themselves as clear and distinct shapes. They present themselves as "vague" shapes that reveal themselves only when the motion stops. (In some cases this can happen if the field is large enough and the speed slow enough for me to fix my eyes upon an object as it moves.) Some form of *fixing* is required to determine the clarity and distinction of the outline shape. Once again the stable and mute object returns as the hidden norm of visualist space significance.

Yet the "weak" or "vague" shape-aspect of the moving object is closer to the many shape-aspects which auditory experience yields in its constant flux. A duration is needed to discriminate shape in this constant motion. Thus if "extended," temporal duration which persists in the flux and motion of sound in time is what appears as the main presentational mode of heard shape-aspects. The much shorter and more "instant" norm of visual stability allows duration to be either overlooked or forgotten and thus apparently to be less important in the visual discrimination of spatial significance.

An edge-shape is "less" than the outline shape, but it is a shape-aspect nonetheless. It is as if the ear had to gradually gain this shape in its durational attention. It is from such temporal considerations that "linear" time metaphors may arise. In this respect auditory shapes seem on one side to be closer to tactile shape discriminations. The blind Indian who concludes that the elephant is like a snake, and who argues with another who thinks the elephant is like a rough wall, is not wrong but inadequate in his "observations" concerning the shape of the elephant. Were he a rigorous feeler of elephants he would not be satisfied with instant apodicity but would withhold his conclusion until he had covered the whole surface of the elephant. So, with listening for shape-aspects it often takes repeated and prolonged listenings until the fullness of the shape appears. This serves no useful purpose in daily affairs when a mere glance will do the same in less time. Thus we fail to hear what may be heard and pass over an existential possibility of listening.

A third variation shows that there is even less absolute difference between sight and sound when size is taken into account. The edge-shape is usually admittedly quite "small." The marble rolling in contact with the box or the die striking the box presents only a small aspect of itself. But visually there is a reversion to a sequential discrimination, too, if the thing is immense. If one stands below the skyscraper, it is unlikely that he will take in the whole at once. He allows his gaze to follow the outline of the building, and the gaze in relation to the vastness becomes a sequential following of the outline-shape. The all-at-onceness does again become possible if distance is increased, as, for example, when I see the whole skyscraper from above while stuck in a traffic pattern in an airplane above Manhattan. Again the comparative reign of the now "middle-size" stable and mute object returns, and the comparative "weakness" and difficulty of auditory shape discrimination returns; but only now it is understood as a matter of relative distancing in space-time. It remains the case that the shape-aspect which is discerned auditorily in its "weakest" possibility is a spatial signification which is limited to a degree within the dimension of hearing.

There is another factor of the hearing of shapes which reveals itself in the "weakness" of hearing the shape of the thing: one that raises the question of *how* the thing is *voiced*. The mute object does not reveal its own voice, it must be given a voice. In the examples listed, for the most part, a voice is given to the object by some other object. One thing is struck by another, one surface contacts another, and in the encounter a voice is given to the thing.

There is clearly a complication in this giving of voice, for there is not one voice, but *two*. I hear not one voice, but at least two in a "*duet*" of things. I hear not only the round shape-aspect of the billiard ball rolling on the table, I also hear the hardness of the table. The "same" roundness is heard when I roll the billiard ball on its felt-covered table, but now I also hear the different texture of the billiard table. True, just as in listening to an actually sung vocal duet, I can focus auditorily on either the tenor or the baritone; but my focal capacity does not blot out the second voice, it merely allows it to recede into a relative background. Thus in listening to the duet of things which lend each other a voice, I also must learn to hear what each offers in the presence of the other. The way in which the mute things gain or are given voices in my traffic with the world is an essential factor in all spatial signification in sound. The voices of things call for further attention.

Although only a massive shift in perspective and understanding will ultimately allow the fullness of auditory spatial significations to emerge, the movement from weaker to stronger possibilities of listening is one that increases our familiarity with such significations. Less strange than the notion of hearing shapes, *we also hear surfaces.* This auditory experience is involved with our ordinary experiences of things.

Who does not recognize the surface in the sound of chalk scratching? I hear footsteps in the hallway. (I can tell if it is Leslie in her heels or Eric in his tennis shoes) or, when the walker steps on the tile its surface produces a characteristic clacking sound of hard heels. Then, the moment the person first steps into the living room the clacking changes to the dull thudding sound of footsteps on the rug.

Surfaces, which are more familiar to us than shapes, must also be heard in terms of a voice being given the things. Just as in the discernment of shape-aspects (and shape-aspects may grade off into surface significations) there is usually a duet of voices in the auditory presentation. Furthermore, there is often more than a surface signification, a signification that grades off at the upper end into an anticipation of hearing interiors. I hear the textural and compositional character of the thing and distinguish easily between the sound of a bell and that of a stick hitting pavement.

Unaccustomed as we are to the language of hearing shapes and surfaces, we may remain unaware of the full possibilities of listening. But the paradigm of acute listening given in the auditory abilities of the blind man often provides clues for subtle possibilities of the ordinarily sighted listener as well. The blind man through his cane embodies his experience through a feeling and a hearing of the world. As Merleau-Ponty has pointed out, he *feels* the walk at the end of his cane. The grass and the sidewalk reveal their surfaces and textures to him *at the end of the cane.* At the same time his tapping which strikes those surfaces gives him an auditory *surface-aspect.* The concrete sidewalk sounds differently than the boardwalk, and in his hearing he knows he has reached such and such a place on his familiar journey.

To be sure, the surfaces heard by the blind man or the ordinary listener are restricted surfaces. They lack *the expanse* which vision with its secret "Cartesian" prejudice for "extension" presents, because the auditory surface is the revelation of an often small region rather than the spreading forth of a vista. But within its narrowness a surface is heard.

But striking a surface and thereby getting a duet of the surface aspects of two things is not the only way in which the mute object is given voice, nor is it only way in which sound reveals surfaces. For the blind man's tapping also gives an often slight but nevertheless detectable voice to things in an *echo. With the experience of echo, auditory space is opened up.* With echo the sense of distance as well as surface is present. And again surface significations anticipate the hearing of interiors. Neither, in the phenomenon of echo, is the lurking temporality of sound far away. The space of sound is "in" its timefulness.

The depth of the well reveals its auditory distance to me as I call into its mouth. And the mountains and canyons reveal their distances to me auditorily as my voice re-sounds in the time that belongs so essentially to all auditory spatial significations. But these distances are still "poorer" than those of sight, though distances nonetheless. This relativity of "poverty" to "wealth" is apparent in the occasional *syncopation* of the visual and auditory appearances of the thing. Such a common experience today may be located in the visual and auditory presentations of a high-flying jet airplane. When I hear the jet I may locate its direction quite accurately by its sound, but when I look I find no jet plane. The sound of the jet trails behind its visual appearance and, by now accustomed to this syncopation, I learn to follow the sound and then look ahead to find the visual presence of the jet.

But as I come to smaller distances the syncopation lessens, and the sight and sounds converge so that ordinarily the sight and sound of the things seem to synthesize in the same place. Yet with careful attention as I stand in the park and listen to the automobiles and trucks rush past, I find that even here there is a slight trailing effect. I close my eyes and *follow* the sound which, on opening my eyes, I find only slightly trails the source as seen. Soon I can detect this trailing with my eyes open. Again in this distance the temporality of sound is implicated.

This often unpracticed and unnoticed form of human echolocation which is spatially significant may also be heightened. For the echo in giving voice to things returns to us with vague shapes and surfaces. The ancient theory of vision that conceived *of a* ray proceeding from the eye to the object and back again is more literally true for the sounding echo's ability to give voice to shapes and surfaces. The blind man, who has learned and listened more acutely than we, produces this auditory "ray" with his clicking cane. Yet anyone who listens well may hear the same.[2]

I repeat the experience of the blind man, carrying with me a clicking device. As I move from the bedroom to the hall a dramatic difference in sounding occurs, and soon, as I navigate blindfolded, I learn to hear the narrowing of the stairs and the approaching closeness of the wall. Like the blind man I learn to perceive auditorily the gross presences of things. But in the relative poverty of human auditory spatiality I miss the presence of the less gross things. I cannot hear the echo which returns from the openbacked Windsor chair, but I do discern the solid wall as a vague presence. Yet in a distance not too far from human experience, I know that the porpoise can auditorily detect the difference of size between two balls through his directed echo abilities, a difference that often escapes even the casual glance of a human.

I listen more intently still. The echo gives me an extremely vague surface presence. I strike it and its surface resounds more fully. Yet even in the weakness of the echo I begin to hear the surface aspects of things. I walk between the Earth Sciences building with its concrete walls along the narrow pathway bounded on the other side by the tall plywood walls fencing off the construction of the new Physics building. In the winter the frozen ground echoes the click of my heels, and I soon know when I have entered the narrowness of the pathway. Once at the other end the sound "opens up" into the more distant echoing of the frozen ground that stretches to the parking lot. But as the days go by and I listen, I soon learn that not only is there a surface presence, not only is there the "opening" and the "narrowing," but there is also a distinctly different echo from the concrete wall and the plywood fence. The surface-aspect only gradually becomes less vague in the sharpening of our listening abilities. In the echo and in the striking of the thing, I hear surfaces as existential possibilities *of* listening.

While there is no question here of exhausting even the relative and often vague "poverty" of shape and surface aspects, the march toward the "richness" within sound must continue. It is with a third spatial signification that this "richness" begins to appear, for, stronger than shapes and more distinct than surfaces, I *hear interiors*. Moreover, it is with the hearing of interiors that the possibilities of listening begin to open the way to those aspects which lie at the horizons of all visualist thinking, because with the hearing of interiors the auditory capacity of making present the *invisible* begins to stand out dramatically. To vision in its ordinary contexts and particularly within the confines of the vicinity of mute and opaque objects, things present themselves with their interiors *hidden.* To see the interior I may have to break up the thing, do violence to it. Yet even these ordinary things often reveal something of their interior being through sound.

A series of painted balls is placed before me. Their lacquer shines, but it conceals the nature of their interiors. I tap first this one, and its dull and unresounding noise reveals it to be of lead or some similar heavy and soft metal. I strike that one, and there is no mistaking the sound of its wooden interior. The third resounds almost like a bell, for its interior is steel or brass. In each case the auditory texture is more than a surface presentation, it is also a threshold to the interior.

I am asked to hang a picture in the living room. Knowing that its weight requires a solid backing, I thump the wall until the hollowness sounding behind the lathed plaster gives way to the thud that marks the location of the stringer into which I may drive my nail. What remained hidden from my eyes is revealed to my ears. The melon reveals its ripeness; the ice its thinness; the cup its half-full contents; the water reservoir, though enclosed, reveals exactly the level of the water inside in the sounding of interiors. Hearing interiors is part of the ordinary signification of sound presence and is ordinarily employed when one wishes to penetrate the invisible. But one may not pay specific attention to this signification as the *hearing* of interiors unless one turns to a listening "to the things themselves."

In the movement from shape-aspects to surfaces to interiors there is a continuum of significations in which the "weakest" existential possibilities of auditory spatial significations emerge.

In all of this listening there is a learning. But that learning is like that of the blind man first being given sight; he does not at first know what he sees. Neither do we know what we hear, although in this case what is to be heard lies within the very familiarity with things in their present but often undiscovered richness. But once we learn to hear spatial significations, the endless ways in which we hear interiors comes to mind. We hear hollows and solids as the interior spatiality of things. We hear the *penetration* of sound into the very depths of things, and we hear again the wisdom of Heraclitus, "The hidden harmony is better than the obvious" (Wheelwright 1966, 79).

In the reverberation of a voice given to things by the striking of one thing by another, in the echo that gives a voice to things, and in the penetration that exceeds the limits of visible space is experienced what is possible for listening. Its presence may occur in the turning of an ear. I go to a concert, and the orchestra plays before me. Suddenly the auditorium is *filled* with music. Here, Baudelaire noted that music gives the idea of space. For now the open space is suddenly and fully present, and the richness of the sound overwhelms our ordinary concerns with things and directions. But even here there lurks just behind us the relative emptiness and openness that the echo reveals. I turn my head sidewise as the music pours forth, and suddenly, dramatically, I hear the echo that lay hidden so long as the orchestra enveloped me with what is sounding before me. And in the echo I hear the interior shape of the auditorium complete even to its upward slant to the rear. The echo opens even filled space, and in hearing there is spatial signification. But let each person listen for himself.

References

Haughawout, Pamela K. 1969. " 'I See' Said the Blind Man." Unpublished paper.

Heaton, John M. 1968. *The Eye: Phenomenology and Psychology of Function and Disorder.* London: Tavistock.

Husserl, Edmund. 1964. *The Phenomenology of Internal Time Consciousness.* Translated by James Churchill. Bloomington: Indiana University Press.

Kierkegaard, Søren. 1959. *Either/Or.* Translated by David F. Swenson. 2 vols. Garden City: Doubleday.

Merleau-Ponty, Maurice. 1968. *The Visible and the Invisible.* Translated by Alphonso Lingis. Evanston, IL: Northwestern University Press.

Straus, Erwin W. 1966. *Phenomenological Psychology.* Translated by Erling Eng. London: Tavistock.

Strawson, P. F. 1971. *Individuals.* London: Methuen.

von Bekesy, Georg. 1960. *Experiments in Hearing.* Translated by E. G. Weaver. New York: McGraw-Hill.

———. 1967. *Sensory Inhibition.* Princeton, NJ: Princeton University Press.

Wheelwright, Philip. 1966. *The Presocratics.* New York: Odyssey Press, 1966.

Chapter 2

The Multistability of Perception

Illusions and Multistable Phenomena: A Phenomenological Deconstruction

In introducing the programmatic and general features of phenomenological inquiry, I have employed typical philosophers' devices. I have analyzed a totality into components and introduced a simplification. The simplification was a visual model of what is global and complex. I am quite aware that the gains in clarity may be offset by losses in richness. What is pushed to the background or, worse, forgotten, may complicate the investigation in the future. In spite of that, I shall take yet another step toward simplification by making the first area for analysis the so-called visual illusions and multistable visual phenomena.

These illusions and multistable phenomena are exceedingly familiar, and so deceptively clear. They are the sort of line drawings that appear in both philosophy and psychology textbooks, and even on placemats in restaurants. Some are two-dimensional drawings that appear to be three-dimensional, such as the Necker cube. In others, straight lines appear to be curved. All the line drawings used will be seen to have some kind of visual effect. It will be the task of a concrete phenomenological analysis to deal descriptively with these effects, illustrating in the process how phenomenological analysis goes beyond what is usually taken for granted.

I point out three things about this group of phenomena: (1) These drawings, all familiar in other texts, are simple line drawings and so, abstract. In comparison to fully etched representational line drawings by

DaVinci or Michelangelo, they are bare line drawings. This clearly is part of their secret—they are from the outset suggestive in their abstractness. (2) This abstractness is a simplification of ordinary phenomena. Although these phenomena, like every visual phenomenon, display all the qualities of a plenum, they do so in a greatly simplified sense. For example, all visual phenomena show extended phenomenal color. But the multi-shaded phenomenon-tree with its greens, browns, blues, and other hues is a complex color plenum compared to a simple white background with black-line drawings. As Eastern sages have long known, the very blankness allows for effects not easily noted in more filled-in configurations. (3) Abstractness and simplicity are conditions for a certain ease in seeing effects, both in their familiar setting and in the more novel phenomenological setting that emerges from deconstruction. Familiarity and strangeness are here bound closely together. They give rise to perceptual games, which are simple, but not without import for the richer and more complex phenomena that can ultimately be analyzed.

I shall introduce one more simplification. In order to display a step-by-step process, often hard to detect in following actual inquiries, I shall introduce an imaginative context and assume subnormal initial perceptions. This device shows the constructive side of the deconstructions, even though normal perceivers begin at a higher level than my imaginary beginners. We are not considering what we already know; we are considering how what we know is phenomenologically constituted.

These simplifications allow the overall experimental movement of phenomenology to be seen as a movement of discovery. This movement begins with what is apparently given, but in the process of variational investigation, the initial given is progressively deconstructed and then reconstructed according to insights derived from the procedure itself. *Epoché* (the suspension of belief in accepted reality-claims) is assumed, and its function as the opening to discovery is shown. Deconstruction occurs by means of variational method, which possibilizes all phenomena in seeking their structures. In this context, *epoché* includes suspension of belief in any causes of the visual effects and positively focuses upon what is and may be seen. Equally, *epoché* excludes abstract generalizations that may apply to the drawings, but fail to account for the specific effects seen (for instance, the generalization that each example is a two-dimensional line drawing, which is true but trivial). Instead, invariants are sought that appear through the variations themselves. With this in mind, the investigation proper may begin.

In what follows I shall name each example according to one of its appearances for ease of reference in subsequent discussion. I shall employ guide pictures, smaller versions of the figures with distinctive features marked on them, in order to avoid elaborate descriptions and make the discussion easier to follow. Neither of these devices is necessary in lectures and teaching, since I can point out what is being discussed. But in the written form, if the reader is to follow the order and be clear about what occurs in each instance, different devices are required. However, the main points should be *seen* in the central, simple line drawings alone. It is important to check your own experience at each step. Of course there are some who will see what is being noted quickly, others less so.

Example One: "The Hallway"

Suppose that there is a group of observers in a room and they are presented with the following drawing (fig. 2.1):

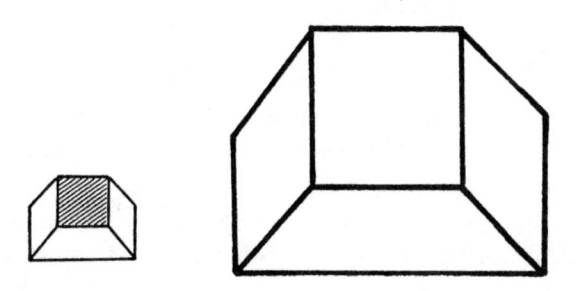

Figure 2.1. The Hallway. Drawing created by Don Ihde.

The interrogator who presents the drawing asks, "What do you see?" The response is divided. One group says that what is seen is a *hallway* (in their interpretation of the guide picture, the shaded area appears to be in the rear of the configuration as perceived three-dimensionally); the other group says that what is seen is a cut-off *pyramid* (in their interpretation of the guide picture, the shaded area appears to be forward, with the sides sloping away from it. In this case, the observer's head position approximates a helicopter pilot overhead). Ordinarily, the hallway group would see the pyramid appearance in a matter of moments, and vice-versa. But for purposes of analysis, suppose the two groups are kept apart and are stubborn in their belief about the appearance they have seen.

Within the framework of the beginning analysis, what each has seen (noema) may be paired with the, at first, literal-minded way in which the initial appearance is taken, with the following result:

> Group H (hallway group) asserts: The figure *is* a hallway appearance.

> Group P (pyramid group) asserts: The figure *is* a pyramid appearance.

Although differing about what "really" appears, each of these groups may be characterized as a variation upon a literal-minded impression of an initial appearance. Strictly speaking, the characterization of the way the initial appearance is taken (literal-mindedly) is a characterization of a total attitude. In this context, the specific noesis (a seeing as) occurs. At this point, both groups are at the same level:

Level I	Noetic Context	Noema
Group H	literal-mindedness	hallway appearance
Group P	literal-mindedness	pyramid appearance

Both groups insist doggedly upon the validity of their reality claim with respect to appearance, but, equally, each sees something different in the configuration. Note carefully, however, that each group actually experiences the noema as claimed; they can fulfill or verify this assertion experientially, thereby offering a certain evidence for their assertion.

This situation is similar to that of the two seers of trees, the "cartesian" and the "druid"; both groups claim metaphysical certitude about what is seen. I shall call this initial certitude, *apodicticity*. To be apodictic means that I can return, again and again, to fulfill the experiential claim concerning seeing the drawing as hallway (or as pyramid).

Simplifying and reducing the example in this fashion may be burdensome to already enlightened people. Most likely, normal observers, already familiar with multi-stable configurations, have long since been able to see both variations alternatively. But viewers able to see either/both aspects are not at the same level of observation as our imaginary groups. They have ascended from the literal-mindedness of the first level by being able to see the alternation (of course, they see either one or the other aspect, but not both simultaneously). Their position, then, may be added as a second level of observational possibility.

I shall term the group that sees either/both aspects Group A' (for ascendant).

Level I	Noetic Context	Noema
Group H	literal-mindedness	hallway appearance
Group P	literal-mindedness	pyramid appearance

Level II		
Group A'	polymorphic-mindedness	hallway and/or pyramid appearances

In spite of the obviousness of the polymorphic observation, several important preliminary points must be clarified even at this elementary level. First, observers of Group A' have exceeded the views of both H and P groups, at least in terms of relative comprehensiveness, since the ability to see both aspects is evidently superior to being able to see only one.

However, there is also an ascent in level. This ascent is indicated by two things that accompany ability to see alternatives. With the ascent in level, the viewer of alternatives does not lose apodicticity; he is able to return to both appearances and thus fulfill or verify the evidence of both. But the *significance* of that apodicticity changes. Once both aspects are possible, it is clear that neither the hallway nor the pyramid appearance can claim absoluteness or exhaustiveness for the possibilities of the thing.

To put this phenomenologically, the noema is now seen to contain two possibilities, and two possibilities as variations are relatively more *adequate* than one. The ascent in level is a move to (relative) *adequacy*, which now assumes a higher significance than mere apodicticity. However, it should not be forgotten that this relative adequacy is attained only through variations upon apodicticity. The fulfillability of the possibility must not be empty.

Simultaneously, the ascent in level also establishes a minimal *irreversible direction* for inquiry. This direction is nontransitive: once the ascent occurs, the observer cannot go back and recapture the naiveté of his previous literal-mindedness. While the variation upon both appearances can at any time be recaptured, the return to the claim that one, and only one, appearance is *the* appearance of the thing is now impossible. This change in the significance of apodicticity is permanent. The direction of inquiry looks thus:

Direction of Inquiry		Noetic Context	Noema	How Held
Level I	H	literal-mindedness	hallway	apodictic only
	P	literal-mindedness	pyramid	apodictic only
Level II	A'	polymorphic-mindedness	hallway/ pyramid	apodictic and adequate

The direction of inquiry in which relative adequacy is more inclusive and yet retains the basic insights of the previous level can move from level I to level II, but once level II is attained, no simple reversion to level I is possible.

In the limited context of this example, in its general import, I am illustrating the first move of *epoché* and the phenomenological reductions. Husserl called for a move from the natural attitude, which is a kind of literal-mindedness imputing to things a presumed set way of being, to what he called the phenomenological attitude, here illustrated by polymorphic-mindedness. By its deliberate search for variations, the phenomenological attitude possibilizes phenomena as the first step toward getting at their genuine possibilities and the invariants inhabiting those possibilities. Potentially, the ascent to polymorphic-mindedness deliberately seeks a particular kind of *richness* within phenomena. But it also carries other implications. These may be seen if a further step-by-step discrimination of what is latent in the current example is undertaken.

Group A' has now discovered that the noema (the line drawing) contains two appearance possibilities. These are genuine apodictical possibilities of the figure. It may be said that they are *noematic* possibilities, inherent in the drawing. But all noematic possibilities are correlated with noetic acts. So far, only the context for those acts has been noted, that is, the type of beliefs surrounding the specific act that give it its implicit metaphysical background. These beliefs are *sedimented*, and the viewer may or may not be able to abandon them.

In this context of "beliefs," each viewer saw the drawing *as* something, either as a hallway or as a pyramid. (It is important to note that in all cases, the viewers saw the configuration *as* something—they did not see a bare figure and then add some significance to it.) This seeing as _____ was instantaneous. The noema appeared primitively as _____. The literal-minded viewers' first look was a naive look. They responded to what first came into view.

Such a response is typical of psychological experiments. Response times are usually limited, and the experiment is deliberately designed to eliminate reflection, critique, or extensive observation. This raises the question what such an experiment reveals, an important question in our phenomenological inquiry. It is possible that an instantaneous glance shows us something basic about perception isolated from so-called higher or lower-level conscious functions. It is equally possible that an instantaneous glance shows only what is most *sedimented* in the noetic context, the context within which perception occurs.

In both cases, the instantaneity of an initial glance must be noted with respect to the noetic act. A simple and often single noematic possibility occurs, which is correlated to the instantaneous glance. Such simplicity was noted earlier to be less than usual, since most viewers would see both alternatives within a very short time. In a psychological experiment with an increased response time, the subject sometimes reports that the figure spontaneously reversed itself. Such three-dimensional reversals are quite common, yet in the standard psychological literature, the noetic act is still noted as a relatively passive stare: If one looks at the figure long enough, "it will reverse itself." If I had not broken down the situation into its phenomenologically ordered components, the standard accounts might seem sufficient.

Remaining within the limits of two noematic possibilities (hallway and/ or pyramid) and within the limits of two noetic contexts (literal-mindedness or polymorphic-mindedness), it is possible to go one step farther. If literal-mindedness occurs when an instantaneous glance is all that is allowed, and if initial polymorphic-mindedness is only characterized as a passive stare, what happens with a more *active* observation? This possibility is easily established by showing that the spontaneous reversal of the two appearance possibilities can occur *at will*.

The *free* variation is easily learned. Return to the hallway/pyramid figure and look at it; when one of its possibilities is fixed, blink your eyes (if necessary, name the other possibility to yourself) and aim for its alternate. Within seconds, or at most minutes, you will find that each of the variations is easily attained at will. This kind of looking is a modification of polymorphic-mindedness. It opens the way to a further stage of the inquiry.

Suppose, now, the figure is presented again, only this time there is a third response from a group of people viewing the drawing (see fig. 2.2). This new group, temporarily placed at the literal-minded level, claims the figure is neither a hallway nor a pyramid, but is a *headless robot*. See this appearance or noematic possibility by returning to the drawing and the guide picture.

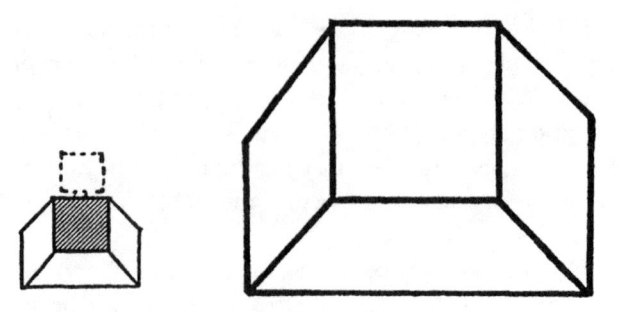

Figure 2.2. The Robot. Drawing created by Don Ihde.

The robot has a body (the shaded area of the guide picture) and is lacking a head (the dotted outline in the guide picture). He needs to support himself with crutches (the outside vertical lines), which run from the ends of his arms (the upper diagonals) to the ground (the bottom horizontal). His legs are the lower diagonals.

Once this appearance or noematic possibility is grasped, fulfilled noetically, a rather dramatic transformation of the first two possibilities occurs. The three-dimensionality of both the hallway and the pyramid appearances is replaced by two-dimensionality; the robot is seen to be flat and standing upright. This noematic possibility is quite distinct from both the others and stands on its own, if the pun can be allowed.

How are we to integrate this possibility with the previous situation? Theoretically, it is possible to take the third group of viewers, Group R (for Robot), as an equally naive group and put them on level I, which now looks like this:

Level I	Noetic Context	Noema
Group H	literal-mindedness	hallway
Group P	literal-mindedness	pyramid
Group R	literal-mindedness	robot

But there is something irregular about this. Empirically, the instantaneous glances allowed in the usual experiments do not produce this third group, or, if it is produced, it is a rare variation.

There seems to be a relative weight or naturalness in the first two variations, which appear three-dimensional, and not in the third, flat variation.

Is it the case that in the first two variations, normal perceptual effects are being noted, while in the third something else is being shown? The question is whether an order of perception or an order of sedimentation is involved. This cannot be answered immediately, but a point can be made about the difference between the usual empirical psychology and a phenomenological psychology. Phenomenologically, the primary question concerns the structure of possibilities, the conditions for such and such empirical occurrences. Its first look is a look for possibilities and their limits, postponing any quick conclusions or generalizations from a straightforward empirical situation. Conversely, its investigation must be concrete, employing actual variations.

At the second level (polymorphic-mindedness), there also occurs a problem of integrating the robot-appearance with the previous situation. Once group A' viewers become aware of the robot appearance, they can add it to the series of variations along with the previous two variations. Group A' now has as its noematic possibilities: hallway and/or pyramid and/or robot. The direction of the inquiry—from apodicticity in each alternative to adequacy—has been maintained. (Once the third variation occurs and is established, one cannot go back to seeing the figure only as one of the three, without the changed significance of apodicticity occurring.)

This new modification of polymorphic-mindedness introduces a new variable and a new question. If what was taken to be *the* appearance of the noema has given way to two alternate appearances, and these have now given way to a third, has the range of noematic possibilities been exhausted? The discovery of a third possibility is not merely a modification of level II; rather, it introduces a new element.

The new element points to the inherent radicalism of variational method. The possibilization of a phenomenon *opens* it to its *topographical* structure. The noema is viewed in terms of an open range of possibilities and these are actively sought noetically. Thus, a special kind of viewing occurs, which looks for what is *not usually* seen. Nevertheless, what is seen is inherent in the given noema. This modification of polymorphic-mindedness is a new noetic context (open), corresponding to the possibilities of the noema (open). The significance of both noema and noesis has been modified.

Although it is possible to interpret this modification of polymorphic-mindedness as either an ascension to a new level or an intensification of the (already attained) ascendance to polymorphy, it certainly radicalizes ordinary viewing. This radicalization can be plotted on the diagram showing development of the method of inquiry as:

Level I	Noetic Context	Noema
Group H	literal-mindedness	Hallway
Group P	literal-mindedness	Pyramid
Group R	literal-mindedness	Robot

Level II		
Group A'	polymorphic-mindedness	alternation hallway/pyramid (ordinary reversals)
Group A^m	polymorphic-mindedness (open possibility search)	alternation hallway/pyramid/ robot/?—(topographical possibilities)

Here, the essential features of the direction of inquiry are preserved, but have been intensified. Group A^m (ascendant modified) has more alternations than Group A' (simple ascendant as opposed to any of the literal-minded groups). *Relative* adequacy is increased and is more comprehensive than in the ordinary alternation. Two certainties are preserved: first, all initial apodicticities are retained, in that they may be experientially recaptured; second, the direction of inquiry is certain, in that the attainment of relative adequacy is intuitively obvious, since each new alternation makes for more adequacy than the previous ones.

How far this procedure can go is not certain. Once alternatives are opened, only the actual investigation can show if closure is possible.

Through the first example, I have illustrated the structure of a phenomenological inquiry, its logic of discovery. If, now, the potential attained at the modified level of Group A^m is read back into the lower stages of observation, one can begin to understand this procedure. The universal level of possibilization ("possibilities precede actualities in an eidetic science") includes, but transcends, all previous levels of the possible. It preserves the validity of each lower level in that it does not lose the ability to re-fulfill each experiential aim. At the same time, it has ascended to a higher level of sight, which transforms the significance of both thing and act of seeing. Suspending beliefs (naive noetic contexts) is needed to open the possibilities of the seen to their topographical features; otherwise, the possibilities are confined to sedimented, ordinary viewing. The radicalized vision of modified polymorphy is not presuppositionless, as some have claimed; it is the attainment of a new and *open* noetic context. In Husserlian language, this attainment occurs in the switch from the natural attitude to the phenomenological attitude.

I have used the hallway example to reveal only the first steps of the essential shape of the phenomenological inquiry. I did not exhaust its possibilities, but though what follows might have been done in terms of this example, I shall turn to a series of similar multistable drawings and begin a more adequate and systematic deconstruction of their initial appearances in order, gradually, to approximate the topography inherent in the noema and the possibility search in a phenomenological noetic context.

Variations upon Deconstruction: Possibilities and Topography

With the general shape of the inquiry determined, the hermeneutic rules of procedure introduced, and one example partially analyzed, it is possible to begin variations upon a group or class of structurally similar drawings. Once again, variational method is employed to open the phenomenon to its topographical features. Such viewing looks for what is potentially there in the noema. This instantaneous glance is far from naive immediacy. It is a view seeking possibilities, as an expert investigator in a field study would look for subtle markings by which to distinguish the creatures he is observing.

For example, bird watching. The viewer notes the general configuration and silhouette of the bird and looks for minor markings, often very difficult to detect. Beginners seldom distinguish among the vast variety of sparrows, yet with visual education they soon learn to detect the different markings of song, chipping, white-throated, and other sparrows. The educated viewer does not create these markings because they are there to be discovered, but—in phenomenological language—he *constitutes* them. He recognizes and fulfills his perceptual intention and so sees the markings as meaningful. So, in what is to follow, vision is educated through a possibility search seeking the constitution of validly fulfillable variations. One aim is to arrive at a more adequate understanding of perception itself through this process.

Example Two: The Curved Line

The first drawing in this series is a very familiar one. The center lines are usually said to appear curved, whereas in reality they are straight and parallel (see fig. 2.3). I call this the curved-line example.

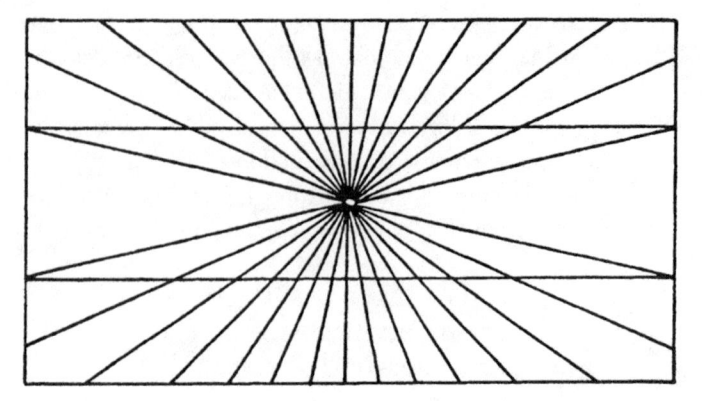

Figure 2.3. The Curved Line Illusion. Drawing created by Don Ihde.

Suppose a situation similar to our previous example, that is, a group of persons observing what appears in the above drawing. Some people view this as an optical illusion, the apparent curve of the central horizontal lines being interpreted as an effect of the configuration.

However, we will begin the noetic-noematic analysis with the preliminary oversimplified situation. One group of observers sees these lines as curved; they take this to be *the* appearance of the phenomenon, and can be identified in the same way as the initially naive groups of example one.

Level I	Noetic Context	Noema
Group C	literal-mindedness	curved lines

Seeing curved lines in the drawing is the normal way in which this configuration is taken in its ordinary noetic context or as sedimented in ordinary beliefs.

Furthermore, this initial appearance seems natural, so much so that a reversal does not spontaneously occur. In this example, too, certain appearance aspects seem to be privileged in normal ordinary perception. But this example is different from our first example, in that there is no reversal. Wherein, then, lies the possibility of alternate literal views or of a second level of viewing with either/both appearances?

Here, we might introduce an artifice to demonstrate that the apparently curved lines are really straight. But, given our method's demand that every variation must be actually experienceable, a problem emerges. Take figure 2.4.

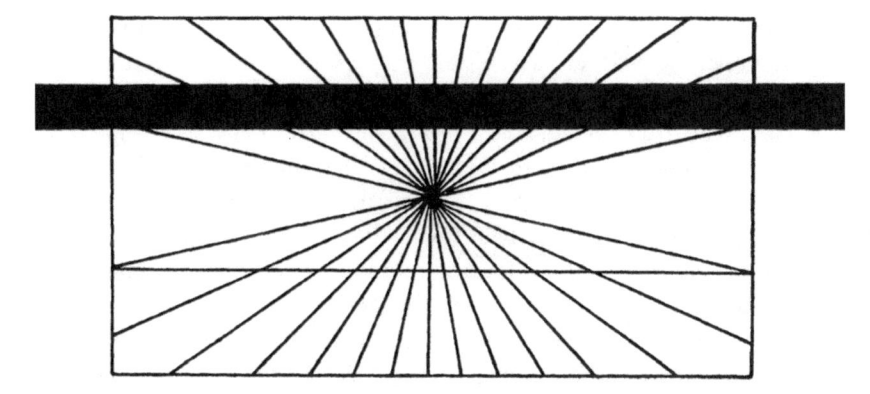

Figure 2.4. Curved Line with Bar. Drawing created by Don Ihde.

Here, an external element has been added in the form of a measuring device laid over the original drawing. Now, noting the clear parallel between the ruler and the horizontal line in the drawing, it is possible to see that the line is not curved. But this is cheating, because it radically alters the original perceptual situation. A curved-line drawing without a ruler is not the same as a curved-line drawing with a ruler. So I shall discard the artifice. If alternate possibilities are to be shown as belonging to the noema, they must be discovered *within the drawing itself*. Return to figure 2.5.

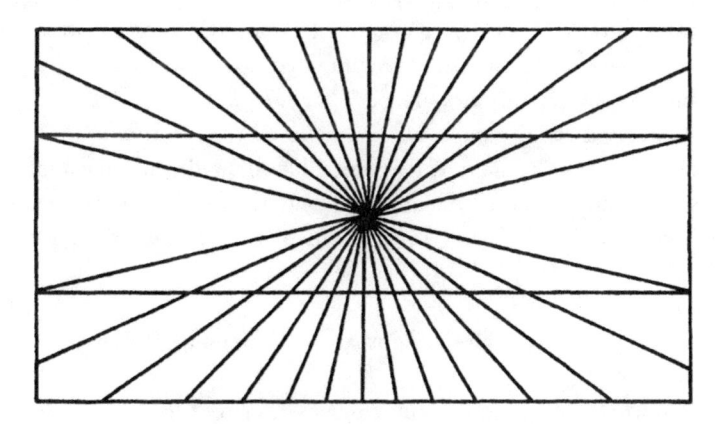

Figure 2.5. Curved Line Revisited. Drawing created by Don Ihde.

This time, look at the drawing in the following way: First, *focus* your gaze intently upon the vertex, where all the diagonal lines converge in the center of the figure. Second, deliberately see the vertex as three-dimensional and in the far distant background, that is, push the vertex back, as it were, until the diagonal lines are seen to lead to infinity. Granting that this takes a certain amount of concentration, subjects usually can do the task quickly, and then the formerly curved horizontal lines appear straight. But this is so only as long as the subject focuses upon the vertex intensely, making the horizontal lines peripheral to the central focus.

Although this kind of seeing seems artificial, we now note an alternative appearance to the drawing. Suppose that in the imaginary situation someone saw the curved-line figure first and simply as a three-dimensional distant vertex. Were that the case, we would immediately have an alternative.

Level I	Noetic Context	Noema
Group C	literal-mindedness	curved lines
Group S^f	literal-mindedness	straight lines (far apex)

Now we can compare the curved-line figure with the hallway figure, in that there is one flat appearance and one three-dimensional appearance.

In order to make the variations completely parallel to those of the hallway example, there would have to be a three-dimensional reversal possible. Return to the figure for a third time. Focus upon the vertex of the diagonal lines, but this time bring the vertex toward the very tip of your nose. That is, reverse (with effort if necessary) the direction of three-dimensionality. It may take more than one try to attain, but once you succeed note that the lines are straight when the focus is maintained upon the vertex of converging lines and the horizontal lines are peripheral.

There is a strict parallel of noematic possibilities with those of the hallway example: there are now two reversible three-dimensional appearances and one flat appearance. The direction and shape of a phenomenological inquiry occurs, with both levels, as it did in the hallway example.

Direction of Inquiry Level I	Noetic Context	Noema
Group C	literal-mindedness	curved lines
Group S^f	literal-mindedness	straight lines (far apex)
Group S^n	literal-mindedness	straight lines (near apex)

Level II

Groups A' to A^m polymorphic-   curved/straight^f/straight^r/$^?$—n
mindedness

While the parallel of two three-dimensional and one two-dimensional possibilities is now demonstrated, there remains a curious difference between what occurred normally and what occurred with difficulty. In fact, in the curved-line example, the empirical order of discovery is the exact reverse of the hallway order. In the hallway example, the first appearances were three-dimensional and easily detected as reversible alternates; the flat, robot appearance came second and perhaps with a little initial resistance. In the curved-line example, the flat appearance with curved lines came first, while the reversible three-dimensional appearances came second and perhaps with difficulty.

Here, the question of whether an order of perception or of sedimentation of beliefs determines the empirical order becomes even more enigmatic. To assert that this inversion of empirical order is perceptual, one would have to maintain that something in the figure naturally determines the order. If one asserts that the empirical order is only the result of sedimented noetic contexts, one would have to show how this order is arbitrary. Unless something appears that distinguishes the way in which the three-dimensional effect belongs inherently to the hallway and not to the curved lines, it is possible there is a tendency to arbitrariness to the empirical order of appearances. But it is clearly too early to conclude this is definitively so.

However, at the phenomenological level of the open search for the topography of the perceptual noema, correlated with the active looking of an open possibility search, it is now the case that the two examples display the same invariant structural features, allowing at least a three-dimensional reversal and a flat appearance, regardless of the empirical order of discovery. But note that something new has been introduced into the inquiry while it is underway. I described seeing the first example as (like) _____ in terms of a kind of story. The drawing appeared as (like) a hallway, or as (like) a pyramid, or as (like) a headless robot. Once the appearance had been named, the observer could easily identify the noematic possibility. This story device, a metaphorical naming, connects the abstract figure with something familiar. The context of familiarity is such that the names are names of ordinary well-known things.

Suppose, when considering the hallway example, we found empirically that all the members of the hallway group were carpenters. They were accustomed to building hallways—in fact, that is almost all they ever did. In this case, the order of appearance would not be surprising at all. Plainly, carpenters who build hallways are likely to see the abstract drawing as something familiar. If we discovered that all those who saw the figure as a cut-off pyramid were pyramid builders, there would be no surprise about the order. At level one, the same goes for the robot group.

Of course, the story device is not neutral. It allows or suggests a certain way for the perceptual act to take shape. But it does so indirectly. One does not have to know what one is doing perceptually to have the appearance coalesce as a hallway or a pyramid or even a robot, once the appearance is recognized as such. One sees spontaneously, once the suggestion is made. This is not to say that the appearances could not have been discovered without names; a story device simply provides an easy and unself-conscious way for the appearances to be noted.

However, while a story device could have been used in the curved-line example, it would have been secondary. For the new element was a set of direct instructions about how to look: focus upon the vertex and then push the lines either forward or back, either toward infinity or to your nose. (As a passing observation, those who are highly successful in pushing the lines to infinity and back again can learn to do this on a continuum, so that the horizontal lines may be seen to be straight, gradually curve, and then re-straighten.) Here, a direct connection can be made with a feature of both the visual field and the noetic act: a ratio between what is focal and what is fringe.

The direct instructions given to the viewer of the curved-line figure deliberately modify the normal or ordinary way the figure is initially taken. In a short glance resulting in the flat variation, the figure is taken as a whole, a totality. The entire figure appears within a wide focus. But under the instructions on how to see, that focus is modified to a narrow focus, with the results already noted.

Phenomenologically interpreted, this is to say that the way in which noetic shape (focusing) is modified is a condition for the possibility of seeing certain shapes rather than other noematic appearances. Nor was this feature absent from the hallway example, though there the role of focus was indirect. In the hallway example, I drew attention by the device of the guide picture, to a particular aspect of it: the central square. In both the hallway and the pyramid possibilities, this indirect focal act played its part. But when the headless robot figure occurred, there was a deliberate, although indirect, drawing of focus away from the central square. Mentioning the absent head, the arms and legs, and the ground on which the robot stood, provided a wide focus and so a flat configuration to the figure. The noetic act functioned to allow this different possibility to emerge.

Here, we move toward resolving the question of whether perceptual or sedimented order controls the way in which empirical orderings occur, but the advance transforms the question. On the one hand, the empirical order clearly contains habitual sediments affecting how a particular configuration is taken. Yet on the other hand, in all appearances, a certain structural feature of perception is operative (in this case, focus—with the added possibility of three-dimensional effect).

For the moment, note two different strategies of interpretation within the investigation. The use of story devices and (metaphorical) naming I shall call a *hermeneutic* strategy. In a hermeneutic strategy, stories and names are used to create an immediate noetic context; they derive their power of suggestion from familiarity or from elements of ordinary experience. The story creates a condition that immediately sediments the perceptual possibility. In untheoretical contexts, this has long been used to let someone see something. Storytellers, mythmakers, novelists, artists, and poets have all used similar means to let something be seen. Plato, at the rise of classical philosophy, often paired a myth or fable with argument or dialectic. Within the context set by the story, experience takes shape.

But, note what happens in terms of the functions and structures of perception in the hermeneutic strategy: neither has to be known (theoretically) nor do directions about how to see occur. Instead, there is a gestalt-event. All at once, the desired effect is achieved; one sees it in an instant. And even if the shift from what was expected or ordinarily taken is dramatic and radically different, once it occurs it is so obvious that one wonders why it was not seen in that way before.

Ultimately, the hermeneutic strategy places its primary emphasis upon language. The symptomatic use of language as a story device illustrates the point. Perception takes shape within and from the power of suggestion of a language-game. It sees according to language. This strategy is the basis of what has become known as *hermeneutic phenomenology*. Historically, the preeminent figure working it out was Martin Heidegger, but Paul Ricoeur also took this direction.

Hermeneutic strategy tends to place its emphasis upon a noematic weighting, although in a somewhat unusual sense. The story lets something be seen; thus, what stands out is the noematic possibility. *How* this occurs, at least in terms of the mechanisms of perception, is less important than *that* it occurs or can occur in the ways open from the topography of the noema. In terms of another long tradition of philosophy, a hermeneutic strategy is more likely to be realist, insofar as it gives a certain precedence to the thereness of the noema. Language is the means, the primary relation, of the condition for the possibility of this discovery.

The second strategy I shall call a *transcendental* strategy. By this I mean that, like transcendental philosophy with its theme centering upon the subject, the way in which perception functions is made thematic. This is the counterpart to the hermeneutic strategy and can be seen to function with a certain *noetic* tendency.

The instructions on how to look rely on a certain knowledge of the mechanisms of perception and on a turn to the subject as active perceiver. This noetic turn stresses the activity of viewing as the condition for the possibility of the object appearing as it does. In its most extreme form, particularly in the work of Edmund Husserl, its tendency is to emphasize the constitution of meaning done by the subject. In terms of the usual traditions of philosophy, this gives the transcendental strategy an idealist emphasis.

Transcendental strategy, however, is also analytic in its procedure. If one knows enough about the structures of perception, ideally, one should be able to deduce, or at least predict, which effects will eventually occur. One may guide viewing from the knowledge of its structural possibilities.

Yet both of these strategies, in spite of a tendency to emphasize one or the other of the correlational foci, point up the same phenomena and utilize variational method to achieve the ultimate understanding of invariance, limits and range of possibilities. In continuing the analysis of multistable examples, I shall utilize both strategies and attempt to demonstrate how a knowledge of invariants is built up in the process.

Chapter 3

What Pragmatism Adds to Phenomenology

On Nonfoundational Phenomenology

This afternoon I would like to talk about a nonfoundational phenomenology. First, I want to set the context in terms of the contemporary state of philosophy, in part in North America, and in part in relation to the traditions which have grown primarily out of the phenomenological tradition with which I am most familiar. In the phenomenological tradition, the three obvious, big masters are Husserl, Merleau-Ponty, and Heidegger, and then a whole series of people who revolve around this particular tradition. Merleau-Ponty takes Husserl in one direction, and Heidegger takes Husserl in a quite different direction. Husserl begins the early 20th century; and Merleau-Ponty and Heidegger go to the mid-20th century, thus we are talking about a tradition in terms of these people and now more than a decade past since the deaths of these giants of the phenomenological tradition. This tradition continues to develop in spite of the fact that philosophy never makes any "progress." (Neither does it stand still, philosophies never die, they just get abandoned, you just simply lose interest in them and you do something else and then they get resuscitated by miraculous resuscitation, revision, or things of that sort, and in part this has been happening.)

There have been two major strands of contemporary philosophy that could be said to be more or less attacks upon the phenomenological tradition. These attacks are not from the outside, as one would expect, let's say from analytic philosophy or positivism or something of that sort. Those would be external attacks and usually the external attacks take rhetorical form like: That's nonsense! That's meaningless! It doesn't mean anything and so you don't read it. You exercise mental hygiene and you simply say that that's nonsense, whether or not you have read it. That sort of attack is not what I am talking about. I am talking about two other strains of attack which are closer to internal attacks. One side, in America, is the recent work of Richard Rorty, and on the other side, for lack of a better term, I'll use the one that has become popular, the post-structuralists, of whom primarily I am going to speak of Foucault, although I could also speak of Derrida in the same tradition. Both of these attacks, if you will, upon the phenomenological tradition, are closer to what I am calling internal attacks. They are attacks from people who carefully read the tradition, and who are trying to push it in another direction.

In America, if one were to talk about mainline, establishment philosophers, who out of the analytic, dominant traditions of American philosophy have looked most closely at elements of so-called Continental philosophy, one would mention Searle and Rorty, with the latter probably being the most eminent of them. He has become a controversial figure.

On the other hand, from the Continent—known in America in a different context, less perhaps in philosophy than in some of the associated disciplines—is the work of Michel Foucault. Foucault was, in fact, a student of Merleau-Ponty and his unspoken antagonist is Merleau-Ponty. Now, what I want to do is relate these two critiques.

A North American Critique of Phenomenology

First let's take a look at Rorty. Rorty's thesis is that philosophy takes different shapes in different times, and it follows certain traditions for a long period of time, and then it abandons or modifies and takes a different shape. This is a very Foucaultian thesis, interestingly enough, where you have what Foucault calls *episteme*. These are eras of organization of knowledge that are discontinuous. They are totally discontinuous in Foucault's sense. The only thing that makes them appear to be continuous is anachronistic reading. We tend to see the previous era as an anticipation of our own and we read anachronistically and selectively, but in fact the system of organization is radically discontinuous. What he is trying to do in what he sometimes calls "archaeologies" or "histories of perception" is show how those different organizations happen.

Now Rorty does not explicitly acknowledge this, but Rorty is doing essentially the same kind of thing with philosophy. He says that early modern philosophy—let's take the Cartesian era, Descartes on through Kant—developed a kind of philosophy that he called foundationalism. Simply put, it is the notion that there must be some absolute foundation for knowledge that is at the base, and then you build up the various sciences and knowledge upon that foundational base. This was part of the project of what we also would call the transcendental traditions. Now, what Rorty says is that this tradition has now come to an end. It is no longer feasible to talk about a foundational philosophy. This model—this "episteme," if you will—exists only as an archaic vestige. The interesting thing is that most people, I think, misunderstood him in the first part of the response to *Philosophy and the Mirror of Nature* (1979) . What he is doing is to address this attack, which has essentially Continental roots, to the current analytic establishment. In fact he is saying that most of analytic philosophy is now at the dinosaurial stage, just about to go extinct, and that only what he calls hermeneutic or edifying philosophy survive this period of extinction. I don't know if the people of Göteborg have heard the news yet or not, but that is Rorty's announcement.

Now, what he puts over on the other side is what he calls hermeneutic or edifying philosophy, and in a sense, it is what I would call a kind of analytic pragmatism. This comes, interestingly enough, from a series of both American, but also primarily Continental traditions. Nietzsche is a strong figure in the background, but even more so, the three figures that Rorty refers to as being nonfoundationalists, as forming the pathway for nonfoundational philosophy, are the late Wittgenstein, the late Heidegger, and John Dewey, all of whom he claims are abandoning the very project of a foundational philosophy. You no longer talk about foundations of knowledge, absolute certainty or anything of that sort. What you have, in Rorty's language, is a series of conversations, and these conversations occur and transform themselves and go on without beginnings, and without endings in any absolute sense. One begins in the middle.

When I first heard of Rorty's books, I was department chairman and did not have much time for extracurricular reading. So, all these conversations were going on and I was believing what the secondary people were saying, and I thought, "Oh, Rorty is another one of those Johnny-come-latelies who has discovered what those of us in the avant-garde of Continental philosophy knew twenty years ago. Thank God somebody is finally catching up with us." If you went back to William Barrett, in *The Illusion of Technique* (1968), he talks about the late Heidegger, the late Wittgenstein, substitutes James for Dewey, and this book appeared several years before the *Mirror of Nature* (1979). He had the same kind of thesis around this different topic. In Continental circles there is nothing new in the attack upon Cartesianism, nor the attack upon foundationalism which had occurred from the beginning of the Phenomenological tradition.

Now, what Rorty does, however, is the following. He is antiphenomenological in a very strange sense. He identifies Husserl simply and purely as a foundationalist, and the same with the early Heidegger, and then dismisses them. He never says another word about this. He simply assumes that Husserl is a foundationalist, that early Heidegger is a foundationalist in *Being and Time* (1972) and that the later Heidegger modifies this. I am not going to go into all of this, but it is the kind of transformation that occurs between early and late Wittgenstein.

Audience Questioner #1: Does Rorty talk about why he came to that conclusion about Husserl?

No. In the *Mirror of Nature* (1979), the main, sustained thesis is a whole series of arguments about why analytic foundationalism is no longer feasible. I do not know how much Husserl Rorty has ever read. This disturbed me when I read it because he is very quick to draw the conclusion, and never discusses it. He just assumes that it is the case. That was part of my own response. One of the things that disturbed me was that phenomenology, which in terms of all the European traditions in America is the strongest tradition, simply gets ignored by Rorty. The whole tradition, which I experienced from 1964 when I graduated from graduate school to 1984 does not exist! My style of philosophy is invisible to Rorty because he does not address it. It is dismissed, so I got disturbed by that, obviously. Because there are nearly a thousand of us who belong to the Society for Phenomenology and Existential Philosophy, some of us have been doing something in that twenty-year period, and it does not exist for Rorty.

Now, I suggested that this is the Rortian thesis, which is very much discussed in America, and what he is trying to do in *Philosophy and the Mirror of Nature* (1979) and *Consequences of Pragmatism* (1982) is to do a kind of outline of his own version of what he calls hermeneutic or edifying philosophy. This philosophy is nonfoundational, and has its roots in pragmatism where there are certain problems that you attack, or you have changes in languages, changes in vocabularies. Rorty remains linguistically oriented in the situation. You do not have beginnings, you begin, if you begin, in midsts, and deal with contexts. (I am not going to talk about Rorty's positive program as such).

A French Critique of Phenomenology

Now, the second attack comes from the European traditions, particularly the French ones in Foucault and Derrida. I am only going to talk very briefly about Foucault. Foucault also explicitly claims that of all the traditions of philosophy, the one he is opposed to the most is phenomenology. He sets himself up as an opponent. Again, interestingly enough, phenomenology is equated to Husserl, and he virtually does not mention either Heidegger or Merleau-Ponty. Everyone who reads Foucault can see that Merleau-Ponty comes very close to doing things that are very similar to the kinds of things that Foucault does. So once again you have this strange identification of phenomenology with Husserl.

Now, Foucault's program is somewhat different than Rorty's in the following sense: *epistemes* are organizations of knowledge and perception which differ in various eras. Very fascinating stuff. If you have not read Foucault, the two most suggestive books to begin with are *The Order of Things* (1969) and *Discipline and Punish* (1975), extraordinarily dramatic works illustrating his notion of *epistemes*. *Epistemes* amount to total organizations of knowledge which are discontinuous. It is a kind of atomization of Heideggerian epochs, if you will. Heidegger talks about epochs of being. They are usually very large, they change from time to time. What Foucault has done is to shrink them down into maybe at most a century long each. They break up in discontinuities of *episteme*.

Let me just use a couple of illustrations very briefly to show what Foucault has done in *The Order of Things* (1969). For example, medieval perception. And he calls it perception. To deal with perception, let's take flowers. How do you perceive a flower in the Middle Ages? Well, first of all, in terms of the medieval sense of perception, you would not even talk about perception. You would talk about seeing and experiencing a flower. But if you do (talk about perception), how do you describe it? Well, what you will have is a multilayer of meaning, which belongs to every—but now see I cannot help but do precisely what Foucault is trying to get us not to do, which is to read anachronistically. I am already reading back into a medieval perception of the flower, a certain classification that we have. So you have, with the medieval bestiaries and so forth, a theological significance, for example. Let's say that the flower is in a heart shape. Now, this significance of the flower will be immediately apprehended as being, for example, the great unifying principle, as the unifying principle of analogy on the basis of certain kinds of analogies and disanalogies. Let's say—and I am fantasizing, but you can see the point—that this is the sacred heart of Jesus. This bleeding heart flower exemplifies Jesus for us. You can see the same thing in the bestiaries. The lion gives birth to cubs that are dead and after three days the male lion comes and breathes the breath of life into the dead cubs, and they are then born as the significance of the resurrection.

Secondly, you have other kinds of symbolic notions to go along with this. What about its health purposes? It is shaped like a heart, it relates to heart diseases, so you give this plant for heart diseases. This is done on the basis of these analogies.

Now, what is the plant? It also has to include its histories, its legends, what is said about it by the doctors, and all that goes into the system of what counts as that flower. You would not dream of saying that you had an adequate description of "flower," unless you included all of these notions.

Let's jump a vast jump now to the Cartesian era, and this one is a mind-boggler to me. By the time you get to the Cartesian era, what have you done? First you have eliminated, and Foucault shows how, most of the levels of meaning. How would you suppose that flower or fauna would be described by the Cartesian era? Again, to oversimplify, what do you do? You abstract into terms of extended shape. Now you might talk about heart shape, to be sure. But you even talk about it as being colorless, so that most of the classifications of flowers have nothing to do with color whatsoever. Why? Because the metaphysical organizational system of the Cartesian era is simply extended substance, and geometrical in that respect. So now colors drop out, and the early classifications have very little to do with colors. Colors have nothing to do with how flowers are related together, and you place them on some vast table of organization based upon shape alone.

Now you jump to the 19th century, and you have yet another system of organization. You no longer use the previous table of classification, but now you are talking about microorganismic processes, so now you classify flowers according to certain kinds of subsystems of organization. What kind of capillary system does it have? What kind of breathing system does it have, or functional systems of this sort? You classify according to a different *episteme*.

So you have three different, if you will, possible histories of describing flowers as the illustration of epistemes. They are discontinuous, they are not overlapping, and they have totally different paradigms if we use the Kuhnian term.

Now, on the larger scale, this has very radical implications, not only for foundational philosophy, because clearly Foucault is denying any such thing as a foundational philosophy, without explicitly saying there is no possibility for foundational philosophy in the Foucaultian scene. What you have at most is a kind of progression of unpredictable changes of paradigms, which one happens after the other, and Foucault in some respects does not give you much of a clue as to how this occurs. There is some hint about how it might occur.

Audience Questioner #1: I thought he objected to even trying to give an account of that.

I think that's right, because in a sense that would now locate him closer to the standard views. But you have a series of discontinuities in which one thing happens after the other, so to speak, in terms of these small epistemic claims. Now of course if you read Husserl as a foundationalist, obviously once again, as in the Rorty situation, you are out of bounds.

One radical implication of Foucault's is, for example, his most notorious claim that man is an invention of the Enlightenment. Man did not exist before the Enlightenment. I want you to see what man means in terms of Foucault's notion of the Enlightenment, you could see that there is a certain amount of sense to his claim because what is a human being with certain kinds of inalienable rights, and so forth, is different from the Socratic inquiry into what the human is—he is pointing out all the differences.

He is a little perverse in one sense. He is a genius in saying if the dominant intellectual history says things are continuous, evolutionary and so forth, then I am going to say they are discontinuous, nonprogressive and so forth. It is like Kazantzakis's Judas: the best way to become a prophet is to know what the swing of pendulum is next, and to jump on that, and you will get the attention you want. I think there is a certain amount of that in Foucault. But he does it with such ingenuity that you cannot help but admire him in the process.

Implications for Phenomenology

What is the result of this? First, as I said, there are these two contemporary attacks upon the phenomenological tradition, both of which, at least with regard to Husserl, regard phenomenology as being one of the last gasps of foundational philosophy, and assumes that Husserl is basically a foundationalist. I am not going to read my chapter in response to Rorty, but my basic argument is that there is at least a superficial sense in which he is right, because certainly the scaffolding that both Husserl and early Heidegger set up looks like foundationalism. They have a basic foundation upon which other things are built, and you need that scaffolding, particularly in Husserl's case, in place. In fact, I have many times argued that Husserl likes his scaffolding so much that he never wants to get rid of it after the building is built. That gets him into all kinds of problems, because what you should do after the building is built is get rid of the scaffolding and let the building stand, but he does not do that. He keeps on using these vestiges of what I would call the transcendental tradition in the work. Whether or not he understood this to be scaffolding or part of the building is, I think, an ambiguous point of interpretation. I am not going to enter that.

But clearly the two attacks have set, at the very least, a new context for interpretation and possibly the preservation of the dying species of phenomenologist. Having been identified with phenomenology, I have only two choices, either I must respond to this, and defend the bastions as they are interpreted, which for reasons that will become clear I cannot do. Or I must become converted to something else, in order not to be abandoned in this situation. I have never liked conversions very much, for all kinds of reasons. So my strategy has been to re-look at the traditions of phenomenology and to discover, in fact, that I do not think phenomenology has ever been foundational. I think that there is a vestigial remnant of the context in which Husserl first found himself, and for that matter, Heidegger first found himself. However, the whole internal trajectory of phenomenology has itself been antifoundationalist. I am going to base that claim primarily upon a kind of Husserlian procedure, rather than the more obvious ones that come out of the later people, although I will refer to them in passing (such as those of later Heidegger and Merleau-Ponty). At the same time, I will not go in the direction of Foucault entirely, because I think the framework of interpretation which occurs in phenomenology continues to make a certain amount of sense as long as it is reinterpreted in light of these criticisms. This is the philosophical context in which theorizing, if you will, about phenomenology as late 20th-century garb, if it still exists, must occur.

Multistability and Variations: An "Empirical" Post-Husserlian Program for Phenomenology

I am going to turn now to a Husserlian-derived or post-Husserlian program, which I think is still phenomenological, and which in fact I think can unite these histories, even with these of the anti-phenomenologists, primarily Foucault, but also in a certain extent Rorty. The program can be billed simply if you will, *multistability* and *variations*. With respect to Husserl, what I am doing is something that can be stated simply, although it is not quite as simple to do as it can be said. Husserl goes through a whole lot of machinery in most of his introductions to phenomenology, talking about shifting from the natural attitude to the phenomenological attitude. He goes through many reductions. How many are there? Some interpreters say three, some say five. He has his own versions of how many reductions you get in order to get the basic phenomenological correlation. That is all, from my point of view, scaffolding. It is a heuristic process to get you to see things in a different way, to get you to understand things in a different way.

According to Husserl's students, including his younger colleague Heidegger, when you actually went to Husserl's lectures, all this written scaffolding became very simple, because apparently in classes he used a lot of concrete examples to establish in a fairly short period of time what was quite tedious and difficult to follow in terms of the published matter. I do not think that is unusual for philosophers. If you read Kant's *Lectures on Ethics* (1785), they are very clear and to the point. If you read his *Prolegomena to Any Future Metaphysics* (1783), you get lost. But, anyway, it seems to me that after you subtract the scaffolding, and you ask, "What are the dynamic factors in a phenomenological analysis that give you some kind of grasp, which is at least unique to a phenomenological point of view and which is suggestive for some line of development philosophically," I think you come down to the notion of variations.

Actually the model behind Husserl's notion of variations was originally out of mathematics. He was doing mathematical variations and applying this model now to different regions of human experience than had been previously applied. Of course, like the mathematicians, he wanted to arrive at what he called an "essence." And, of course, the Husserlian essence is very difficult, because it has a long history of terminological use in other parts of philosophy, and he is radically modifying that use. I am not going to get into that debate, but simply to point out that is where he comes from.

Now I want to look very briefly at what happened in the history of philosophy in Husserl's sense. He wanted to say that fantasy variations were the privileged ones. Here are two examples. Again, this is like a mathematical procedure. You are a sort of mathematician, you are sitting there and theorizing in your head, with a piece of chalk on a blackboard and so forth. Whatever else is going on you have already abstracted from. That is the model of fantasy variations. But let's get a little more concrete. What about imagination and perception? Husserl, at least through his mid-career—I am told by my very rigorous Husserlian scholar colleague that I should correct myself with some of the later obscure manuscripts which have never been published, and I cannot read Husserl's shorthand—and in the early days, Husserl, in the *Ideas* (1962), for example, clearly adapted the empiricist false prejudice that (a) imagination is dependent upon perception, and (b) that it is at most a kind of copy, perhaps a poor copy of perception. Now, he does not take that latter point quite as badly as the empiricist tradition does, because he is clearly emphasizing imaginative variations. When you say that you can do your imaginative variations in order to find the essences of things, so you might think that—you do not have to be a geographer to discover the terrain of the world—if you are a Husserlian. In that sense, you could imagine it, and recreate it in terms of fantasy of variations. Well, it seems to me that one of the things that happens—it already happened in Merleau-Ponty—is the recognition, pretty quickly, that this set of assumptions is, in fact, phenomenologically false. Imagination does not replicate perception, and furthermore, it is not a poor variation of perception. It is quite *different* from perception. And thank God for that, because otherwise we would be doing the same thing as perception. There is a variation, or variational difference, between imagination and perception which a phenomenological investigation will show.

Let me give you a few examples to show how this is the case. In terms of concrete analysis, perception is an interesting case. I think it is probably referred to in one of my earlier books, *Listening and Voice* (1976), where I am doing a phenomenology of sound. Let's take visual perception. Visual perception has a number of phenomenological structural features which are well-known in certain aspects of psychology as well. Let's take just a couple of simple examples. In this case, I am going to have a little bee. You can see a bee. It is a figure against the background. In this case, if there was an actual bee that I could get here, I could have it buzzing around in front of you. Now, let's vary this a little bit. Let's say that I am sitting on a chair, and the bee is in front of me. How do I see the bee perceptually? Well, it is flying around there, and as long as it is in the range of my focal central vision, it is fine. I can get that bee pretty clearly as he is flying around. If he starts going to the side, I can turn my head and follow him to a certain extent. But now, what if he flies around to the back of my head? I no longer see him. I cannot get him visually. Of course we know that part of the reason is you have a limitation to your field of vision, and once the object exceeds that horizon of visual field, you no longer get it visually. This is not to say that you are not getting the bee, because you can still hear him back there buzzing, and he disturbs you precisely because you cannot see him. Maybe you will turn around and keep your eye on him, but in terms of this reference system, the shape of your visual field is such that you have this horizonal limitation to the visual field.

Let's switch briefly to the imaginative field. Give yourself an imaginative bee in front of your face. A green one, if you like, with yellow wings. Now fly it to the back of your head. This is a difficult variation. But all I have to do to show you that imagination is not isomorphic with perception is to show you in some sense that the *shape* of the imaginable compared to the perceptual field is different. So all I have to do to prove, phenomenologically, that these are not isomorphic is to, for example, get the bee further than the limits of the visual field, before it disappears. That is all I would have to do, to prove my case. That would be the weak form.

I have done this with many, many people, in classes and so forth, and a good many people, in fact the majority of people will say: Yes, I can imaginatively see it in the back of my head without disembodying myself. So the imaginative visual field does not have the same shape as the perceptual visual field. And if that is the case, the realm of imagination is not isomorphic and not dependent upon the visual perceptual field.

I can give 40 different, other kinds of variations to establish the same kind of point. But all you are trying to do, is to show that the imaginative domain is not a representation of the perceptual one. That is phenomenologically false. But now, if that is the case, you have a much more difficult path, than Husserl thought in terms of doing variations. One of the things that follows from this is to say that if you are going to deal with perceptual phenomena, you cannot use imaginary phenomena, except in a very limited and critical sense, to substitute for perceptional ones. You are going to have to deal with the perceptual phenomena within its proper.

What if you go to other kinds of things? We will jump to some of those other kinds of things shortly. But first, what follows from this is that you obviously have to make phenomenology a much more—I'll use it in quotes—a much more "empirical" system, than the kind of system that Husserl pretended it was in the early stages. It is still strictly Husserlian by the way. You are after the phenomena as they present themselves within the limits of the way which they are presenting themselves. That is very strictly Husserlian. But now you have a qualification that the investigations are going to have to be in terms of the various regions and domains that are actually being investigated. So if you are doing cultural investigation, it is going to have to be in terms of that kind of domain; if you are doing perceptual ones, in terms of that kind of domain and so on. That is point one.

Let me go to another Husserlian example. In the "Origins of Geometry" (in *Crisis*, 1970), Husserl has a very interesting set of notions. (Which by the way, I think are derived from Heidegger. The Heideggerian critique of Husserl in *Being and Time*, follows this in the following way.) For Husserl, the primary objects of the world are very much like the objects that occur in empiricist and rationalist philosophy (i.e., you have material objects, you have an X, such that it has such and such predicates). He defined these in terms of a *sensory plenum*. That is, any ordinary object has color, extension, and so forth: all these multiple qualities form the ordinary things of life experience. It is from that, that you move on. What Heidegger did was to say: No, that is an epistemological prejudice. This is not, in fact, what appears—in this foundationalist sense—first at all. Instead there are actions, and actions constitute pragmata, and pragmata are not yet objects. Only when the action cycle is broken do you get an object. So in the famous tool analysis, when you are using a hammer, the hammer is not an object, it is a means of accomplishing a certain kind of action. Only when that project is broken can it become a hammer which has such and such obstinate, persistent, and lasting qualities. Therefore epistemology is, if you will, derivative from action theory, and objects are derivative, if you will, from actional contexts.

What Husserl does, in some sense, in the later work in *The Crisis of European Sciences and Transcendental Phenomenology* (1970), is to learn some of those lessons, and when he gets to the "Origins of Geometry," he begins to do the same kind of thing. He still insists on having his objects, but now they belong to a sensory plena, and he is still quasi-foundationalist. I agree with that criticism. And these are the ordinary objects of the world.

Now, how do you get geometry? Geometry is the increasing abstraction and idealization of certain parts of that sensory plena. So what is basic are these ordinary objects: tables, chairs, and all the stuff that philosophers always talk about. The first thing is a practice of abstraction. You can see, I hope, coming around the corner, our friend Foucault here. The first process is a process of abstraction. Instead of regarding all these objects simply as what they are—ordinary sense, material concrete, dynamic, all these things—you abstract into shapes: these kinds of shapes and those kinds of shapes, but not just abstraction (sub) one, but abstraction (sub) two. Not just shapes, but particular shapes, because Husserl is talking about originating geometry. And geometry at the beginning is totally incapable of dealing with any shape whatsoever. It has to deal with certain particular shapes, and historically of course you are talking about things like measuring practices. For example, the Egyptians remeasuring, resurveying their fields to get nice square or equilateral or whatever shaped fields you want. So, out of all the possible realm of shape, you chose certain ones to analyze geometrically, and of course you know that triangles and lines and squares and so forth arose. But there is nothing in the things themselves that says you have to go that direction. That is not only an abstraction of a shape from a thing, but it is an abstraction of certain particular shapes, and in fact Husserl uses carpentry examples.

Now, what if you use my "religion," sailing. You did not begin with a geometry of compound curves: that came much, much later. You could not deal with such curves in early geometry (formally) and yet the praxis of building ships, with compound curves is as ancient as it can be. You did not need a geometry to do the praxis, and the praxis of geometry did not come out of it. So you have a kind of double abstraction that is going on in Husserl's notion of the derivation of geometry.

What you have in Husserl himself is something that is very like this history of geometry. This is part of his selective prejudice, if you will, for the way in which he is going to analyze things. This is one of the reasons why I think he ends up characterizing his work as a transcendental idealism. After all, the ideal entities are the trajectory, the goal toward which he is going. I think it causes him all kinds of problems.

I have shown a couple of things about Husserl. But the practice of Husserl is, of course, to establish his essences on the basis of variational theory, even though I will later qualify the interpretation of this variational theory.

Macroperception and Microperception:
Merleau-Ponty's Extensions to Husserlian Phenomenology

There are a number of post-Husserlian directions which could have been followed. The Husserlian one was in a certain direction toward ideality. The Merleau-Pontian one was in a different direction toward what has become known as the *lived body*. And what Merleau-Ponty (1962) makes as his basic notion is perception—now broadened to include what I would call two dimensions. You have, in Merleau-Ponty, both what I call macro- and microperception. Microperception is sensory perception, as we might ordinarily call it. Macro-perception is what we might call cultural perception.

The reason I introduced this terminology is I think one of the things that our background friend Foucault is doing, is blurring the difference between micro- and macroperceptions. What he has seen in the history of ideas is precisely that there was never any kind of analysis of microperception until a certain era. It did not exist until a certain era. And he, of course, is predicting in one respect that it is going to disappear in another era. So he is not going to reify this kind of notion.

I am making this qualification so that we can see how it relates to the various periods in the thoughts of perception. Macro-perception, in Merleau-Ponty, is cultural perception. Let's turn briefly to a couple of Merleau-Ponty's examples.

If Husserl restored the notion of multidimensionality to perception, what Merleau-Ponty does is to establish a certain set of variabilities with respect to perception. His favorite examples are aesthetic ones.

Take the phenomenon of depth in paintings. One of the most familiar of the traditions concerning depth is so-called Renaissance perspective. Here was a certain geometrical kind of depth created by a system of lines converging on some central point of infinity near the center of the painting. Indeed, a system of lines and a grid were often used in constructing the abstract upon which the painting was to occur.

If, now, we apply the phenomenological correlation of what is seen with how it is seen, of relating the object of sight with bodily positionality, how is depth perceived in the Renaissance examples?

Well, interestingly enough, you can get it from any number of viewing positions. In other words, you have a variability of position with respect to this painting, in terms of where you can get the depth. It is not neutral, but it is flexible with respect to the position from which you see.

To be sure, it is a little less flexible in terms of closeness or farness. In fact, if you deal with *trompe l'oeil* work, which are sorts of technical expertise of such perspectives, you have to be fairly close to get the real "illusion" of depth. That is one of the reasons why you have the key-hole pictured in *trompe l'oeil* works. Merleau-Ponty was interesting because what he showed was that this is certainly not the only way in which depth is created. His favorite painter, of course, was Cézanne. The way Cézanne got depth was through slabs of color. Very different from Renaissance perspective, and no one can deny that you can get an extreme sense of depth in Cézanne's *Mont Sainte-Victoire* and his still lifes, and so forth.

But now you do not have the same kind of positionality that you do with Renaissance painting. For example, you have to have a much closer preferred range of position to get the full length of depth accordingly. Let me use an example that Merleau-Ponty does not use, but which I experienced very dramatically recently: Monet. Monet is an absolute genius of portraying depth. In the Marmottan Museum, there is a painting of a landscape across an English river. If you are close up, all you get are little smears of color. Totally indistinct. And the farther back you get from it, the deeper the scene gets. By the time you are (20 meters) back from it, it has so much depth that you feel it is almost like a *trompe l'oeil*. In this case, you have a different position, in order to get the depth, than you do in the Cézanne case, or than you do in the Renaissance case.

What you are doing is showing the polyvalent ways in which depth can be represented in terms of painting. What Merleau-Ponty is saying is this is a mini-example of the relationship between what I am calling macro- and microculture of our perception. In fact, this has to be learned, just the way everything else is learned. The first people who saw the impressionists—it is quite interesting if you go to the Rouen Cathedral series. People, at first, were not able to see it as a cathedral, and I think the very first time I saw it, I had a little trouble myself. But once seen the cathedral becomes so vivid and dynamic in this case, that I cannot *not* see it as a cathedral, at this point. Yet people did not see it as a cathedral at first. It had to be culturally acquired. Merleau-Ponty is saying that perception, in this sense, is culturally framed in terms of a kind of polyvalence. Well, this is doing variations, and showing that there is no one way. There are multiple ways of portraying depth, artistically. And each of these constitutes its kind of mini-world, which is, in many cases, not commensurate with each one of these systems.

Audience Questioner #1: How important is this distinction, because I have a little bit of difficulty in seeing this as cultural and not individual?

Oh, at this point, I would not even want to differentiate between individual and cultural. That is because we are *in* this culture. I think it might be, at this particular level, difficult to show what the connection is, but there are certain other examples in which I can show how it is more culturally located. You see, one of the things that is happening culturally is something we have already acquired—and this is part of my argument—this is why I think nonfoundational philosophy is already part of the standard rubric by which we deal with things. If I use a Foucaultian interpretation, we have already moved into a relativistic era, so that we are very familiar with doing precisely this kind of thing. That is a reflection of our life situation, which is not a reflection of life situations of, say, other kinds of cultures. To show this I would have to move to a much more radically different culture than the one which we have. There would clearly be individual variations among people, depending on educational background and so forth within our culture.

Let me use two familiar and almost standard examples. Take snow, a good old-fashioned one. For, the average North American farmer, how many kinds of snows are there?

Audience Questioner #1: Four.

Yes, two or three or four. For the Eskimo, how many kinds of snow are there?

Audience Questioner #2: Seventeen.

At least. So you have a finer perceptual discrimination, because the life praxis that you have in these kinds of situations demands distinctions.

Same thing with camels, you know the standard example of camels in Arabia where they have all these different names for camels, compared to our one smelly, dromedary beast. So that is one set of distinctions. That would be cultural in relation to the praxis, the experiential praxis of the thing.

The second one will be more dramatic. If you do it between more radical cultures, there is a study—I do not remember exactly how it goes—in which people are given abstract drawings. There are certain cultures that initially see them as two-dimensional, whereas we do not—if you are talking about an empirical progression and so forth. They habitually see them differently, in a different order than we do. That would be a cultural rather than individual difference. This would be sedimentation, in a sense, with that multistable example that I gave before. It probably relates to some fairly subtle practices.

In most cases, one of the interesting things about this is that most of the cultures that see it (as two-dimensional) are not accustomed to having pictures of this sort. So you must initiate a whole picture-learning process. Or in the early days of film in certain cultures, when the person walks off the screen, the viewers wondered where did he go, because the notion of framing is not yet learned in that respect. That would be the kind of thing that I would point to for those cultural differences.

From Foundational to Postfoundational Philosophy

So now we have got variations that encompass both micro- and macroperceptions. I should probably have these in quotes, because if you are Foucault, it might be the case that we will either move in one direction or another to interrelate or not interrelate micro and macroperception.

I think, by the way, what I am talking about is not something that belongs simply to Foucault. It is not something that belongs simply to Merleau-Ponty. It is something really that is close to what Foucault is calling an "episteme" of the age. I think it also belongs to Kuhn; the paradigm shifts in *The Structure of Scientific Revolutions* (1970) have to do with historical eras in science. What Kuhn is doing, in effect, is a kind of Wittgensteinian seeing in terms of Gestalts. So it is not something that belongs to one tradition. It is something that belongs, somehow, to the very situation that we are in historically, culturally at this point. What the philosopher is doing, if you will, is not doing foundational philosophy, but is doing a kind of critical reflection upon what has happened to our "episteme," our perception of the time.

Audience Questioner #1: It is hard say whether it is our perception or those of the people who spend a lot of time reading this.

Right. That is fair enough. "Our" means community in a Kuhnian sense too.

I earlier talked about Husserl and multidimensionality being restored. Husserl's multidimensionality is not the restoration, in a full sense, of the medieval multidimensionality. The loss of theological significance in the thing is pretty much of an irretrievable loss. In fact, Husserl is very much of a vestige, a transitional figure to me, between the foundational and postfoundational periods. There are always vestiges of foundationalism, and you can see over and over again between what he will allow to perception over against apperception: the famous perception of other persons in Cartesian meditation number five, for example: apperceiving on the basis of a body, another person and so forth. That is a vestige of empiricism which remains in Husserl, and which Merleau-Ponty sought to simply eliminate. You do not perceive bodies, to which you apperceive minds. You perceive other persons, from which you can subtract to get objective bodies, for example. It works quite the other way in Merleau-Ponty compared to Husserl. So, the multidimensionality that Husserl returns is not a premodern one. It is a different one, in keeping with the microperceptual structures discovered by phenomenology.

The other direction you get, say, in Wittgenstein. Wittgenstein really does not have anything like sensory perceptions, as far as I can see. He is really talking about cultural perceptions, seeing as cultural perception. Sensory perception is, in some sense, included within cultural perception. It is taken for granted almost, within. In this respect, he is much closer to Foucault.

In sum, we have got Husserl establishing multidimensionality, Merleau-Ponty coming along with polyvalence, and I could illustrate the same movement in Heidegger on a more historical basis.

Multistability and Structures: Recent Developments in Phenomenology

What makes for a new phenomenology comes from combining the notions of multistability and multidimensionality with the notion of structures. In an earlier work, *Experimental Phenomenology* (1977), I undertook some of this work. My illustrations were of visual multistability, such as that found in drawings as of the famous "Necker cubes" (fig. 3.1).

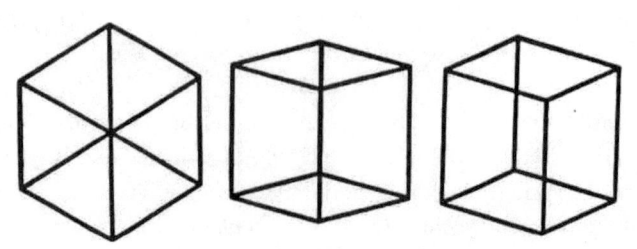

Figure 3.1. Necker Cubes. Drawing created by Don Ihde.

Here is a series of such drawings and we are familiar with the usual interpretations. The cube series is an example of multistable perceptual objects in which there are a number of possibilities. The standard interpretation recognizes a bistable, three-dimensional effect. Either the cube is tilted "downwards" towards the viewer (note that the apparent position of the viewer also subtly shifts slightly to "below" or "above" in a small degree). I shall call the two standard three-dimensional (3D) effects 3Da and 3Db.

But the standard view does not exhaust the multistable possibilities. The "cube" may also be seen as two-dimensional. This is easy in the case of the "cubes" which is a hexagon in which the central cross point must be taken to be an overlapping set of lines to constitute three-dimensionality. But, once discovered, it is equally easy to get to a two-dimensional effect from the most familiar cube drawing. Just imagine the central parallelogram of the drawing to be the body of an insect; the outer hexagonal outline to be a hole; and the lines from the "body" to the "hole" to be the "legs": voilà, two-dimensionality. (In lectures, I sometimes use a styrofoam construction instead of a drawing to make the next point easier to grasp). But one need not stop with 3Da, 3Db, and 2D. Once having the "insect," imagine pulling its body-part (the parallelogram) outward from being bound flatly to the "hole" while imagining the "legs" to be elastic (in my styrofoam construction the "legs" are elastics) and pull the "body" toward you. In this case you get a new three-dimensional effect which is not at all that of a cube—and if you are now getting swiftly phenomenological, you will know that you can do a reversal of this effect as well by pushing the "body" in the opposite direction.

So, now you have not bistability, but pentastability: 3Da, 3Db, 2D, 3Dc, and 3Dd. In *Experimental Phenomenology* (1977) I show that these same five variations belong to each of the drawings of the series (nor do these variant exhaust the series). Indeed, once the perceptual secret of variations is learned, one may see that there is a virtually mathematical progression which could be employed to predict the number of possible variations. For each part of the drawing, the three-dimensional effects may be projected forward or backwards, with other configurations between the parts (all of which can also be constructed in three-dimensional material objects as well) to make for a very large series of possibilities.

This is, in one sense, simply an illustration of the precedence of an *eidetic* "science" in Husserl's sense, over an empirical one, for the possibility structure in his foundationalist sense "founds" any merely empirical science. But in the nonfoundationalist context the investigation of possibility structures may not be taken as more than the exploration of a full range of variants.[1]

I have argued that this particular set of phenomena are linear arbitrary. In some sense, the most you have to begin with are perhaps constituted habitualities about the way you tend to see them. Once you break through those, you do not necessarily get them in that order at all. The order is arbitrary. It is like a kind of mathematization of perception.

You do not want to only raise the question of what kinds of variations are possible. This seems to me to be standard, straightforward, Husserlian-derived phenomenology, but it is clearly not foundational in the old sense. It is foundational perhaps in a secondary sense because you clearly are getting at essential knowledge, that is, what is the range of possibilities, and the range of possibilities will have different kinds of structures. So, you may have a linear and arbitrary structure, or graded structures, or hierarchical structures. You may have dependent, founded relationships. All those things are possible in terms of the particular kind of phenomena you are investigating, and you do not know what those are, until you do the actual investigation. But this is clearly not the traditional kind of foundationalism because there is no particular or favored kind of foundation.

Questions about the Recent Developments

Audience Questioner #3: This is the essential question point. Do you mean that these are different structures, or do you mean that the variation has a structure of its own? I think that here you are referring to the structure of variation. Those are the different possible modes of seeing the figure. Here again. There again.

In one sense you are perfectly right. If we go back to the simple Husserlian correlation, I think the framework of the Husserlian interpretation is the interaction in the interaction. If you say it is the structure of variations, you are locating it here. Let me use an illustration. In the Husserlian intentionality diagram—or the Heideggerian version of it—you have got equal primordiality in which there are three distinguishable, but nonseparable elements: myself, the world, and the interaction between us, if you want to put it in a simple language. Now, interestingly enough, you are opting for, I think, a perfectly good option. And I suppose I would be willing to opt for that as well as anything else, because all that I am saying is, if you change something at the world pole, you also change it at the relational and I pole. It is a relativistic structure in that respect.

Let's take two illustrations to get at this. In one sense, I do not care which is emphasized. That is what I am saying. In the early days of radio, with the earphones on the radio, von Békésy had an interesting set of problems when he was doing his early works on communications theory (he was a Nobel laureate for his work on sensory inhibition). Some people—when they heard the sound—asked themselves, where did I hear the sound coming from? And some people heard the sound as if the orchestra were in front of them. They were used to that, and that sounded right. (I have not used the Walkman, so I do not know if you get the same multistable illusion or not.) Other people would get the sound as if it were in back of them. And of course, you can go into an exploration of why that occurred, a 180-degree error in terms of the directionality of sound. And you could go into a whole series about time differences and wave shapes and so forth to give a physicalistic interpretation. But, nevertheless, you heard sound in front, or you heard it in back, or some of them heard it in the middle of the head. From my point of view, it does not matter, all three of these possibilities are equal. They are fine. The only person who is going to be disturbed about it is the person who wants to have it appear in front or in the middle of the head instead of behind. It makes no difference to me where it is located in that sense, because the phenomenon of the dimensionality, the spatial dimensionality of the sound, has that multistable feature.

Let me use another illustration. It comes out of the nothing of a Foucaultian *episteme*. Take color. Where does color occur in the different *episteme*? If you have eye and world, taken in epistemic fashion, prior to the modern era, with the exception of certain Greeks, where did you have color? Color existed in the object. It was one of the properties of the object. What do you do when you reach the Cartesian era? You displaced color. Color no longer belongs to the object, it belongs to the eye. Color is a secondary property, invented by the subject. It does not belong to the real object at all. The real object is only extended material substance: it does not have any colors. What you have done is displaced color. But phenomenologically speaking, where is the color? You have to, in some sense, choose this. Color is an experiential function of what happens to me in my seeing of the world. It is both in the world, and in me, it is between us in that respect. So my long-term answer is: I do not care where you put it. Because the phenomenon stands out as one of the invariants in the situation, it does not make any difference to me whether you chose this *episteme*, this *episteme*, or this one, as long as one can descriptively analyze the phenomenon as such.

Audience Questioner #3: But my question was whether the structure refers to the variation, like if you look at this figure, that you pointed out all the different instances, and the way in which you arrive at those differences, that is the structure.

Yes, I agree with that. I think what is happening in this emergence is a kind of relativism, one that I want to address, because people get upset about relativism.

It is not a relativism in the subjectivist sense at all, it is a relativity rather than a relativism, because in relation to any of these kinds of situations, you have got to have some relationship between, let's say, a fixed point and a dynamic situation. What is relative is that you can move this around. But you have to have to have both in relation.

Take the famous Einstein situation. We have three moving trains. How can you tell which train is going in which direction? It goes back to the business of stabilization of positions. If you are in an airplane above, you can see which train is moving in relation to which other train from that position. But if you occupy one of the train positions, you can have an equal account about any number of possibilities, but not all possibilities. If one train is moving in one direction, and the other train is moving slowly in another direction, that is a possibility, but you have to take into account the position of the observer. What you could describe absolutely, if you will, is the relative speed of that train in relation to this train, from the point of view of the observer. There is nothing unabsolutistic about that. But what is unabsolutistic about it, is taking that particular position. You do not have to take that position, from this particular example, if you could do variations accordingly. So it is not relativistic in that sense; it is a relativity, which always takes into account some position. Of course the problem is that the all positions are also multistable.

Audience Questioner #3: But the structure, in this case, relates to what is fixed and what is dynamic?

Yes. Right, and you can change from one to the other.

Audience Questioner #3: Yes, so your claim is that the variation has a structure, and that is a position which in some extent transcends the two positions.

That is what I am claiming. It is my assertion that this goes beyond Husserl and Merleau-Ponty in terms of what I am calling variational method and the model of stability.

Audience Questioner #1: So structural analysis of a variation is identifying what the fixed point is, what the changes of the points are, and their relation. And just one more terminological point. The word "variation" usually brings to my mind, considering several cases, and then looking at comparisons between them. As I understand your use of the term, it refers to a single instance.

No. It could refer to a single instance, but it does not have to refer to a single instance. This is again a very Husserlian point. You can have an adequate intuition of a single instance, if you are lucky. But you are usually not lucky.

Audience Questioner #1: What I am thinking of is the way I have just described the structure of variation. My understanding is that you take a single case, you look for the point of view that you are taking and the changes. But all the analysis is internal to the case. But if you take two cases, now you can make an analysis across the two cases, talking about relations in one and not in the other. In fact, mathematics is built to express certain concepts with this one set of axioms, which cannot be expressed with another, even though it is about the same thing, so to speak.

Exactly. I think in this respect it does parallel exactly that kind of procedure, except now it is using perceptual, and other kinds of experiential aspects, rather than simply mathematical ones.

Audience Questioner #1: But there is a different kind of analysis, it would seem to me for analyzing a single instance versus comparing across instances. I am not clear whether you want to make that distinction.

Yes, I do want to make that distinction.

Audience Questioner #1: What would some of the differences be?

Let me go back to the *Experimental Phenomenology* (1977) to make one other point about this. Let's go back to our headless robot creatures. Variation one is a 3D rear. The simplified story I told to show what I am after, and which I think responds to your point is this. If you have a tribe of people who by virtue of their culture, and their experience, know only my 3D rear variation, they take that to be the appearance of this particular thing. I call these people literal-minded. We still have lots of those types around in the world who say this is the reality of that. And then you have, imaginatively, another tribe of people who have 3D forward. And they are also literal-minded. You can have a tremendous cultural conflict between these two mini-tribes because they will each argue for their case accordingly. It is certainly true experientially that they are both getting (their particular interpretations) strongly, purely, repeatedly, and so on. But as soon as you get a person who comes along who says, "Oh, there are in fact two aspects to this," then you have destroyed the possibility of remaining literal-minded, because you have changed the meaning of what happened. So now you can call it polymorphic-minded, if you will, in this case.

This is a different level of analysis because it changes two frameworks. Now look what happens. You do not lose the experiential fulfillability of the single instance, but you change its meaning. It now becomes weaker than the point of view which is polymorphic. It is retainable, but it becomes weaker.

I suppose that you could use religious examples. Once, in the interpretation of biblical literature, you discover that there are additional features to it, you do not necessarily lose the sense of religiousity, but certainly you lose the sense of fundamentalistic religiousity. You cannot go back once you have found the critical perspective.

Of course when you get to variation three, or you get 2D, you have extended this yet another step. You have not necessarily ascended the entire level, but you can argue—once you get the two variations in—that you have now moved towards a greater sense of adequacy. This is an old Husserlian notion. The single instance could be what he would call *apodictic*. It is repeatedly fulfilled. It is certain in that sense. But apodicticity turns out to be a weaker term. *Certainty* turns out to be weaker than *adequacy*. Adequacy is when you start in a polymorphic direction, and my argument is, once you start in that direction, it is a nontransitive direction. You cannot go back. You only go in this direction, and you do not know how far you can go along that trajectory.

Audience Questioner #1: In this example, you still keep your analysis of an individual case. What is changing is your interpretation of the significance of that analysis from the individual case. But you have not really said anything about whether there is going to be an additional analysis of adequacy. In fact, I think in your writing, you talked about how you can try to give an analysis of all the possible structures that appeared, and maybe you could say some more about that.

Yes, I would say, however, that analysis is going to be strange in another respect. There is a strong sense in which even weakened apodicticity remains an important part of what is going on. What can you say once you have taken this irreversible route? You can begin to derive, retrospectively, as it were, differentiations between certain structures: linear structures, hierarchical structures, graded structures, and so on.

Only after you are so far down the road, you can also do a second thing. You can do a kind of heuristic interpretive rule-making. And I want to be very careful and qualify that to say that this is always a kind of post facto. I am very Kuhnian in that respect. Once you have seen it, then you can make your rules and use them in terms of the next step.

For example, I say: I know that if the cube drawings have these kinds of appearances, I will be able to heuristically predict that a new set, that you give me, will have the same structural features. All I have to do is to investigate, and find out if they do. Now, if they do not, I am going to be very surprised. So far I have not found any such figure I cannot analyze. Maybe someday I will find an abstract figure of the same sort that I cannot analyze, but I will wager almost anything that it is not the case.

That is a heuristic rule. It gives me a guidance into what to look for next on the basis of what has been attained here. I think that is a second thing that is derived from this analysis. It gives me some insight where I can look next, or what I can do next.

In one sense I am returning phenomenology to its originally *investigative* aim, its "empirical" aim. And in that regard I am probably closer in spirit to what Husserl wanted. But in a second sense, by emphasizing investigation through honing the variational method, I think I have moved away from the Husserlian framework towards a kind of phenomenological relativity. These two senses lie behind what I have characterized as nonfoundational phenomenology, some results of which will appear in my book *Consequences of Phenomenology* (1986).

References

Barrett, William. 1968. *The Illusion of Technique*. New York: Anchor.
Foucault, Michel. 1969. *The Order of Things*. New York: Harper and Row.

————. 1975. *Discipline and Punish*. New York: Harper and Row.

Heidegger, Martin. 1972. *Being and Time*. London: Harper and Row.

Husserl, Edmund. 1962. *Ideas*. New York: Collier.

————. 1970. *The Crisis of European Sciences and Transcendental Phenomenology*. Evanston, IL: Northwestern Press.

Ihde, Don. 1976. *Listening and Voice: A Phenomenology of Sound*. Athens: Ohio University Press.

————. 1977. *Experimental Phenomenology*. New York: Putnam.

————. 1986. *Consequences of Phenomenology*. New York: State University of New York Press.

Kant, Immanuel. 1783. *Prolegomena zu einer jeden kunftigen Metaphysik die als Wisenschaft wirt auftreten können*.

————. 1785. *Grundlegung zur Metaphysik der Sitten*.

Kuhn, Thomas S. 1970. *The Structure of Scientific Revolutions*. Chicago: Chicago University Press.

Merleau-Ponty, Maurice. 1962. *Phenomenology of Perception*. Translate by Colin Smith. London: Routledge and Kegan Paul.

Rorty, Richard. 1979. *Philosophy and the Mirror of Nature*. Princeton, NJ: Princeton University Press.

————. 1982. *The Consequences of Pragmatism*. Minneapolis: University of Minnesota Press.

Chapter 4

What Phenomenology Adds
to Pragmatism

Response to Rorty

Were philosophers to have Academy Awards, and were reviews in the major journals and such national papers as the *New York Times* or the *New York Review of Books* nominations, then Richard Rorty's *Philosophy and the Mirror of Nature* would certainly have won the 1981 "Oscar." For that book, with only one contender,[1] surely received more attention than any other philosophy book for that year. The interest spread to Stony Brook, too, for in the fall term Patrick Heelan organized an informal seminar of some twenty faculty and doctoral students to read *Mirror*. I was, due to a schedule conflict, unable to attend, but soon began to hear discussion in the hallways.

Bits and pieces began to take shape: "Rorty's decreed the end of analytic philosophy." "Rorty's turning Continental." "Rorty thinks that the three greatest philosophers of the twentieth century are the late Heidegger, the late Wittgenstein, and Dewey." And, most often from my ACE[2] colleagues, "Well, it's about time somebody from the establishment discovered what we've known all along about the state of contemporary philosophy." Still, I had not read the book.

Then the reviews appeared, repeating in many ways the above conversation in more academic prose. Some believed that with Rorty's strong emergent interest in *hermeneutics* evidenced in *Mirror* and with the repeated references to Heidegger, Habermas, Apel, and Derrida, that he had, after all, opened the way to Continental philosophy. Moreover, his scathing critique of what some identified as analytic philosophy was unmistakable.

But it was not until early 1983 that my own schedule allowed me to read the book. (I admit to some grudging reluctance at first, since, from the secondary information, I, too, believed that Rorty was a Johnny-come-lately to a perspective on contemporary philosophy which many of us arrived at fifteen or even twenty years ago! His choices of representative giants had even been anticipated in print far earlier, and again by William Barrett in his 1978 the *Illusion of Technique*. Barrett chose two of the same individuals, although he substituted William James for John Dewey, but with much the same impact. Moreover, the death of Modern Philosophy, i.e., the "foundationalism" of the Cartesian sort, had been decried by virtually every "classical" phenomenologist in all three varieties—transcendental, existential, and hermeneutic.) Thus when I began to read, I received something of a surprise. First, the bulk of the book did not really have so much to do with anything like either a conversion to Continental philosophy or a deathknell for analytic philosophy that the secondary interpretations seemed to emphasize. Rather, there was a reworking of a total perspective upon contemporary philosophy which in its most penetrating sense did not even distinguish clearly between analytic and Continental forms.

Surely, Rorty's audience remained the AE, and his rhetorical and conceptual style remained clearly within those boundaries. I have already remarked upon the obvious invisibility of those who in ACE were already quite aware of the negative side of *Mirror*'s result. *Mirror* was, in the thrust of its attack, a kind of Kuhnian shift, a change of model or of categories by which one could interpret contemporary philosophy. It was a shift which, negatively, did claim that the end of Modern philosophy, insofar as it was a systematic, epistemological-metaphysical enterprise which constructed itself upon a *foundational* base, was no longer tenable. This was not exactly news to many of us—but what Rorty did further, was to develop the thesis such that his emergent perspective which differentiated between foundational and edifying *hermeneutic* philosophy cut across *both* analytic and Continental fronts. Its result was, on one level, to undercut much of what had been taken for granted as differences between the two styles of philosophy, and replace it with another.

I also found that I had to take Rorty at his word concerning what he was doing, for this attack upon foundationalism in both analytic and Continental groups, arose primarily from within analytic philosophy itself, from its more pragmatist sources. Rorty claims:

> I began to read the work of Wilfred Sellers. Sellers' attack on
> the Myth of the Given seemed to me to render doubtful the

assumptions behind most of modern philosophy. Still later, I began to take Quine's skeptical approach to the language-fact distinction seriously, and to try to combine Quine's point of view with Sellers'. Since then I have been trying to isolate more of the assumptions behind the problematic of modern philosophy, in the hope of generalizing and extending Sellers' and Quine's criticisms of traditional empiricism. (Rorty 1979, xiii)

The hard core of *Mirror* is exactly that. From within the larger analytic movement, Rorty has taken a pragmatist, antifoundationalist stance and argued that all forms of analytic foundationalism are untenable. This is clearly a severe attack, for it implies that the "science model" held by the early Positivists and retained through most foundationalistic philosophy must go.[3] And, if correct in a Kuhnian sense, this would also mean that a lot of what is taken as "normal" problems for analytic philosophy would not so much be solved or reworked, but simply disappear, become "uninteresting." This would be the case for much of the so-called body-mind problem as well. Philosophies are, of course, rarely responsive to refutations. Historically they either tend to be abandoned rather than die of rebuttal, or, more likely to undergo resuscitation by revision. Thus analytic foundationalism has in very recent years, re-emerged as the New Realism Rorty refers to in his later *Consequences of Pragmatism*.

The internal attack, addressed to analytic philosophers, clearly arose primarily from Rorty's own readings of that tradition and its problems. This is clear both in the form and substance of the attack. Nor does it repeat the same kind of criticism made much earlier by both Husserl and Heidegger. Indeed, I suspect Rorty would think of their attacks as still within the foundationalist framework since it is possible to interpret Husserl as rejecting Cartesianism on behalf of transcendental idealism, and Heidegger's destruction of the history of ontology which covers over a more ancient and favored ontology as yet another foundation.

However, in one crucial way Rorty's attack does function like the earlier attacks of Husserl and Heidegger. Rorty's "paradigm shift," which resituates a perspective upon contemporary philosophy, is in practice something like the deliberate tactic of a "paradigm shift" employed by Husserl well before Kuhn. Husserl made such a shift of perspective an essential and deliberate part of phenomenology itself. This tactic, buried for some beneath his intricate machinery, is nevertheless exactly a purposeful shift of perspective. I refer, of course, to what Husserl called the shift from the "natural attitude" to the "phenomenological attitude," a shift which was both deliberate and fundamental for a different kind of "seeing" to occur. The elaborate steps of the *reductions* which appear in most of Husserl's works are the parts which go together, or better, the set of hermeneutic instructions which tell how to perform the shift. Unfortunately, too many interpreters simply either got lost in the intricacies, or worse, read Husserl literally.[4] What was important was to be able to experience what is seen differently, which, once attained make the instructions either intuitive or unnecessary. (In a sense, Kuhn describes what happens in a shift, but how it happens remains for him, largely unconscious. Husserl attempts to make shifting a deliberate procedure, a phenomenological rationality.)

The common perceptual model between Husserl and Kuhn is the *gestalt shift*.[5] Gestalt shifts, either those which take ambiguous figures and grounds, are mini-illustrations of changes of perspective. Paradigm shifts or the shift from the "natural" to "phenomenological" attitudes are, of course, both more sweeping and more fundamental. But it is clear that Husserl made this phenomenon one central element of his own phenomenology. Rorty, with reference to the field of contemporary philosophies, practices such a shift.

Rorty's shift at its highest level of abstraction, divided contemporary philosophies into those which are *foundational* and those which are non-foundational. Then, within those which are nonfoundational, he discerns a certain pattern which he terms alternatively *edifying* or *hermeneutic*. This is to say that the negative side of *Mirror*, the attack upon foundationalisms, is matched by a positive side, the development of a generic hermeneutic or edifying philosophy. It is here that the representative giants take their shape and role: late Heidegger, late Wittgenstein, and Dewey.

If each reject systematic, hierarchical, structural and foundational philosophical erections, particularly those of the Modern or Cartesian and Kantian sorts, then what is to be accepted also has need of clarification. For Rorty, language remains the guiding thread. In his preface, he indicates that what unites all he learned from the variety of his teachers and philosophical education was, "I treated them all as saying the same thing: that a 'philosophical problem' was a product of the unconscious adoption of assumptions built into the vocabulary on which the problem was stated—assumptions which were to be questioned before the problem itself was taken seriously" (Rorty 1979, xiii). This remains a central philosophical axiom with Rorty and he has even produced a lexicon translating the main terms of *Being and Time* into vocabulary language.[6]

Language is a twentieth-century obsession for philosophers, both Anglo-American and Euro-American. But what is common to the giants

Rorty cites is what may be called a certain *horizontalization* of language. Negatively, this is a rejection of hierarchies to language and specifically a rejection of the language/metalanguage developments of foundationalists. Wittgenstein, in the later "language game" and *Philosophical Investigations* period, of the three giants, was the most explicit about this connection. But the movement in Heidegger from the apparent structural foundationalism of *Being and Time* to the increasing hermeneutic horizontalization in his late work follows the same trajectory.[7] And, Dewey, although not directly concerned with language in the same way, unites both in his deconstruction of the sciences into human problem-solving activities which, again, are horizontalized.

Horizontalization implies negatively that there are no privileged language games, no disciplines, no privileged activities. There are only appropriate or inappropriate contexts and a diversity of fields. What seems as a kind of democratic anarchism here, Rorty seems to apply to the very profession of philosophy. It must take its place as one type of conversation among others. Certainly, philosophers can no longer pretend to be either the overarching thinkers of the past, nor cultural mandarins with a higher authority. They can be *edifying* in the sense of a moral concern: "The only point on which I would insist is that philosopher's moral concern should be with continuing the conversation of the West, rather than with insisting upon a place for the traditional problems of modern philosophy within that conversation" (Rorty 1979, 394). This is a modest view of philosophy, an essentially communicative, interpretative one. But Rorty intends it to be positive within this modest position; his choice of edifying is deliberate: "Since 'education' sounds a bit too flat, and *Bildung* a bit too foreign, I shall use 'edification' to stand for this project of finding new, better, more interesting, more fruitful ways of speaking" (Rorty 1979, 360).

Moreover, edification is essentially a *hermeneutic* activity; in its new mode:

> The attempt to edify (ourselves or others) may consist in the hermeneutic activity of making connections between our own culture and some exotic culture or historical period, or between our own discipline and another discipline which seems to pursue incommensurable aims. . . . It may . . . consist in the "poetic" activity of thinking up . . . new aims, new words, or new disciplines, followed by, so to speak, the inverse of hermeneutics: the attempt to reinterpret our familiar surroundings in the unfa-

> miliar terms of our new inventions. In either case, the activity is . . . edifying without being constructive. (Rorty 1979, 360)

One can detect here a certain "Continental" drift but still within Rorty's constant of philosophy as linguistic activity. In the sense and to the extent that Rorty has seriously absorbed the lessons of Continental hermeneutics, the notion that he has turned toward Continental philosophy is only partly true. But insofar as hermeneutics is itself interpreted as linguistic activity, having to do with a kind of Rortean language game in which new vocabularies are made, discovered, or whatever, the hermeneutics takes its place within his overall analytically conceived notion of vocabularies and their unconscious assumptions. In this respect *Mirror* is simultaneously analytic and Continental, while, equally simultaneously, it shifts the divisions of philosophy such that the high level difference between foundational and edifying philosophies cuts across both analytic and Continental philosophies.

Although I must return to the notion of horizontalization which plays such a crucial role in edifying philosophy, it may be of interest to see how the Rortean scheme might divide the field of phenomenological philosophies. (I note that Rorty effectively ignores phenomenology as such and in the few instances where it appears at all it is identified, particularly in Husserl, as belonging to the foundational enterprise.) Is all phenomenology foundational? Or are there foundational and edifying phenomenologies?

Is Phenomenology Edifying?

At first reading Husserl would indeed seem to be a prime candidate for a foundational philosopher. The architectonic of his overall work clearly has that structure. He is avowedly a *transcendental* thinker, placing himself in that respect in the Descartes-Kant traditions. He "founds" his architecture upon the ground of transcendental (inter) subjectivity. He claims phenomenology as a "rigorous science," a new science, thus not unlike Positivism, incorporates his version of a science model into his method. Moreover, most of his critics and scholarly interpreters have taken him that way.

Yet there are elements within his philosophy which are implicitly antifoundational. Interestingly, these are the devices of method which make phenomenology function as a *horizontalization* of phenomena. These are sometimes hard to detect because they must be isolated from within the apparent structure or foundationalist architectonic which scaffolds Husserl's edifice.

In this respect, the strategy of Husserl's *Cartesian Meditations* is peculiarly instructive. Husserl simply accepts the structure or framework of Descartes' project, but then, step by step, he replaces each element with a radically different result. For example, on the surface it seems that Husserl's and Descartes' foundations are the same: the *ego cogito*. But in the end, they are not. Descartes' ego is (a) the self-enclosed subject, (b) worldless except by inference or "geometric method," (c) a subject without object, etc. Husserl's transformations replace each of these elements: (a) the (phenomenologically present) world is equiprimordial with and strictly correlated with the ego; (b) the ego, thereby is *not* self-enclosed, and in fact is reached only by way of the world; (c) there is no subject without an object, nor object without a subject. In short, the whole building has been replaced and the scaffolding alone gives the semblance of a Modern philosophy.

Another, more Rortean way of phrasing this is to say that what remains after Husserl's *deconstruction* of Descartes is a new vocabulary. It is the vocabulary of the correlation of noema and noesis, of I and World, of *correlations a priori*. Moreover, if anything is "given" in Husserl, it is what is always "given" in the Rortean scheme, *some vocabulary*. Then, within this vocabulary, there are grammars of movement about how one may go in one or the other direction. I shall contend below that these hermeneutic rules at the core are the *variational methods* which derive from phenomenology. But in both senses, what remains of the Modern project is scaffolding—the problem is that Husserl was always proud of his scaffolding! This is evidenced by the vast amount of his publications which had to do with describing it in the multiple ways he did (how many reductions are there? how many ways of getting to the ego? to the phenomenological world? etc.).

It has always been my contention—admittedly disputed by many literal-minded Husserlians—that Husserl's method was heuristic. Over and over again, he adopted the terminology and the structures of Modern philosophy in both its Cartesian and Kantian forms, but in each case he reworked elements and structure, such that they no longer meant what they originally did. In his last work, *The Crisis*, he noted that what he called the "phenomenological attitude" was not, as often earlier described, a device of method, but a permanent acquisition of the philosopher. But, as I shall contend below, the result of this shift of perspective—even in Husserl—is one which is fundamentally nonfoundational.

What is hard to decide concerning Husserl himself, is how much of the radicality implied in his work was discerned by him. There is no doubt that he was wedded to his terminology of "transcendental idealism," even if transcendental meant for him something radically different than in the Modern traditions and even if idealism also was intended to be different from all other idealisms. But there is little doubt that the two founders of variant phenomenologies both rejected transcendentalism (and at least by implication, foundationalism) and saw some of the more radical implications of Husserl's methodology.

Merleau-Ponty's existential phenomenology early claimed that the implication of phenomenology was not transcendental, with all the hubris of a total and self-contained system, but existential. Moreover, the late Merleau-Ponty reworks the I-World vocabulary of Husserl into an ongoing set of interrogations as in *The Visible and the Invisible*. No foundational standpoint is possible, but the polymorphy of the intertwining with its open-ended implications replace the Husserlian scaffold entirely. Similarly, Heidegger's *hermeneutic* phenomenology, although retaining a vestigial foundationalism in *Being and Time* moves as Rorty himself has seen and appreciated in a nonfoundational direction. The expositional debate in this case revolves around Heidegger's gradual dropping of phenomenological terminology, the ambiguity as to whether what he *does* remains essentially phenomenological, and whether or not the moves into his later terminology arose through the very implications of the earlier, more explicit phenomenology.[8] Thus, if phenomenology can still be used to characterize its existential and hermeneutic versions, in its later phase it becomes ever more explicitly nonfoundational.

This excursus, however, is historical and interpretive. What is needed here to expose the edifying and nonfoundational aspects of a phenomenology, is to show from its very core what motivates this possibility. Is phenomenology edifying?

To accomplish this move I shall take two seemingly contrary steps. First, I shall try to show that the edifying or hermeneutic thrust of phenomenology can be found in or arise from essentially *Husserlian* notions (while not denying that nonfoundationalism becomes more explicit in the post-Husserlians). And, second, I shall focus not just upon the concepts and explicit claims about phenomenology, but upon philosophical *praxis*. What dissolves the apparent contrariness between the "antique" and contemporary situations itself arises from the tradition of philosophical practice. In all of this I admit that not all *scholars* of the tradition would agree with me—but I suspect most practitioners would.

The scaffolding of which Husserl was so proud, here interpreted as a set of hermeneutic rules for proceeding, included an emphasis upon experience and evidence. Experience must be actual or fulfillable; evidence is *intuitive* (that which in fulfillment is present). What most standard interpreters have taken this to mean is that whatever is intuitively *given* provides not only some kind of foundation, but belongs to the myth of the given (and Sellers was one of these interpreters). But this interpretation misses entirely the role of evidence and intuition as hermeneutic rule *to discover something else.*

For even within the heart of Husserl's explicit set of procedures, he follows what, to my mind, was a mistaken heuristic which confuses issues. The surface or explicit steps, when put into *practice*, reveal *not* the above kind of given, but something quite contrary. *For phenomenology, intuitions are constituted, not given.* Only already constituted intuitions are "given" *within an already sedimented context.* When taken as "evidence" the evidence is strictly indexical (thus hermeneutic). What the scaffolding allows one to get at is the relationality between experience and contexts or fields. All experience is *context-relative.* Here, at a most basic level is a first clue to phenomenological *horizontalization.*

A second "device of method," as scaffolding, is the shift from "natural" to "phenomenological" attitudes. This deliberate shift, however, is not a shift on similar levels. The phenomenological attitude is *the access to context relativity.* That is why it must become permanent—it is the now attained vocabulary wherein both the new ways of saying can be undertaken, and the inverse hermeneutic of reinterpreting the world can be performed.

This is to say, that once the structure or the field of possible contexts is open, the phenomenological attitude provides the way to explore possibilities which is its field. In the edifying sense, this means the exploration is one which seeks *to find what intuitions can be constituted.* And it is here that variational method emerges as the central driving engine of an edifying phenomenology.

Before following that implication, I would like to take note of one fundamental difference between Rorty's "hermeneutic" and a phenomenological one. Phenomenology has never been simply a linguistic philosophy, although there are strong variants, with Husserl and Merleau-Ponty as "perceptualists," with Heidegger and Gadamer more "linguistic." In all cases what counts as language is always experiential, and even better, perceptual. In phenomenology it would better be termed a language-perception pairing. Even Dasein is concretely bodily-spatial and *Being and Time* rather than merely talking about how one is to perform a phenomenology, undertakes it in relation to human spatio-temporality. But it must be understood that perception here means the perception of phenomenology ("lived body," "lifeworld," "time consciousness," etc.) and not that of Cartesian or Modern, neo-Cartesian physicalism.

In fact, the language-perception link in phenomenology is also tied to both context relativity and to variational praxis. For example, if intuitions are constituted, not simply given, then the task for variational explorations will be to find out in what situations, contexts, cultures, times, "x" intuition can or will occur. This is an actual hermeneutic investigative practice. Husserl preferred fantasy or imaginative variations, and through his midcareer he simply accepted the (false) empiricist assumption that imaginative variations could simply substitute for any other kind, particularly the perceptual kind (imagination duplicates perception). This practice, modelled upon mathematical procedures, but also a favored shortcut for abstract and writing-room-bound philosophers, was thought to be sufficient to yield the invariants of the field of possibilities.

The post-Husserlians challenged this limitation upon variational investigation and developed other practices. Merleau-Ponty (with a different cognate disciplinary background, psychology) noted that perceptual variations could not be substituted for by imaginative ones and developed this inquiry most thoroughly in *The Phenomenology of Perception*. And in a more historical vein, Heidegger took the same tack with respect to his variant *epochs* of Being. Here were historical—and in the case of his "Conversation with a Japanese"—cultural variants. In each case the phenomenology involved, particularly as a practice, uncovered or hermeneutically exposed how the "intuitions" are constituted by the context. For Merleau-Ponty the context is motile bodily position with its interaction with the environment; for Heidegger it is the constellation of historical beliefs which sediment and account for some (then or now) current state of affairs.

Interestingly, this insight and practice derived directly from Husserlian phenomenology, and in spite of explicit rejections of "phenomenology," continues brilliantly as *praxis* in both Derrida and Foucault! For example, in an earlier piece, I have shown how Derrida is doing something of a standard phenomenology of reading in his play upon margins and the like:

> Take a text: If one views a text (perceptually) it usually appears first as a writing that is centered on the page, surrounded by margins; but the focal center is clearly the bulk of what is written. Then, if one reads the text, what usually emerges as focal is what the text is about, however complex that may be, as indeed any text usually is. What does Derrida do with a text? Posed in the

way I have indicated, he immediately decenters what seems to be focal and immediate. His focus is radically shifted to titles, signatures, margins, borders, divisions, etc. In short, he draws our attention to features that are there, but are usually taken at most as background, secondary, or unimportant features.

In a sense this is a highly "phenomenological" technique. For example, in an analysis of perception, phenomenologists like to point out that while what stands out (figures) are usually most obvious because they are the referenda of our usual perceptions, all figures take their position upon a background that is equally present and that constitutes the field of perceivability. In short, this move "decenters" focal perception so as to attend to taken-for-granted but important fringe features. Similarly, to point out that all perceptions include not only manifest surfaces, but latent "backsides," is to "decenter" at least the usual interpretations *of* perception. I am suggesting that his device—perhaps taken to Nietzschean excess—is a familiar ploy of Derrida. Indeed, one can see, once the operation is known, how to follow along with such deconstructions. (Is there a Derrida text that addresses itself to the empty background of the page? If not, there ought to be.) (Ihde 1982, 8–9; also reprinted in Ihde 1983)

Foucault, too, continues the *praxis* of some distinctly phenomenological habits even while linking phenomenology with Husserl and opposing it. His unmentioned teacher, Merleau-Ponty,[9] remains his subterranean mentor. Foucault does histories of perception, as in the *Birth of the Clinic.* That is to say, he traces the radically different ways things are seen in correlation to the different practices of an epoch. (Foucault has a miniversion of Heidegger's epochs of Being, but Foucault's are smaller, more discrete, more rapid in change.) This praxis which continues the development of contexts of language-perception is perhaps most dramatic in *The Order of Things.* Not only is his outline a subtle response to Merleau-Ponty (who claimed there could be language about language, but not painting about painting. *The Order of Things* begins with Velázquez's "Las Meninas," a painting about painting.) The intricate pairing of experience in language-perception is precisely the forte of Foucault who may have adopted unconsciously the phenomenological vocabulary, but who *does* what I would term a kind of subterranean edifying phenomenology.

If, in Husserl's case, phenomenological edification is implicit, and if in the post-Husserlians the scaffolding and transcendentalism places existential and hermeneutic phenomenology on at least a nonfoundationalist trajectory, and if what unites this development is a certain *praxis* which may be either implicit or explicit, then, once having taken Rorty's shift seriously, the question can become more explicitly that of the possibility of an edifying phenomenology.

There may be a quibble here: why call it phenomenology? Heidegger ceased to use the term. Derrida and Foucault, by linking it to the foundationalist, transcendental enterprise, reject it. But if the trajectory I have outlined obtains and the *praxis* underlies what I would term an extension of variational method is the case, then there is at least more continuity than is usually allowed. Perhaps what is suggested is a *new* version of phenomenology, an edifying phenomenology which has at its core precisely that hermeneutic and inverse hermeneutic performance which freely explores what Rorty calls the exotic (histories, cultures, disciplines, etc.).

A (New) Phenomenology Which Edifies

What I have been suggesting is that phenomenology in the late twentieth-century—whether it is called that or not—has had a more and more nonfoundationalist trajectory. Both the explicit transcendentalism of Husserl and the vestigial foundationalism of an early Heidegger have given way to the now dominant strains of hermeneutics and post-structuralist enterprises of the present. Perhaps out of some unsuspected conservativism, but more likely out of a philosophical preference of actional (as opposed to epistemological) analyses, I have chosen to retain the ancestral name.

The same applies to the occasions represented here. A collection by its very nature is not a systematic or accumulative development. It, rather, presents themes, examples, applications which are united only by the unconscious vocabulary of a style—but which better would be seen as vectors along a trajectory. Husserl taught us *how* to do phenomenology. But once learned, the scaffolding which allowed one to get at the edifice is seen to be secondary. Different values about what is central emerge.

Phenomenological *praxis,* I would contend, revolves around an active variational inquiry. Variational inquiry may be imaginative, perceptual, historical-cultural, or interdisciplinary. It thus looks a bit like what Rorty calls hermeneutic or edifying philosophy. But also because variational inquiry is linked to a sense of experience with its language-perception pairing, there is a certain perspective upon things. It is a perspective which links a sense of *position* (from where are things seen? the vestiges of noesis) *referred* to a *context* or field (what and how do the things, old or new, appear?). But it is not tied to any preferred foundation. To be sure, if variations are perceptual, there is the privilege of the lived or motile body, but if they are intersubjective, imaginative or whatever else, that focus is displaced.

Here in this brief response to Rorty, I attempt to follow the trajectory of a nonfoundationalist phenomenology. I take this to be following the *consequences* of phenomenology. The thematic occurrences of plays upon gestalts, the development of cultural-historical variants upon the perceptions of technology, the cases of context relative phenomena are examples along this line. Yet, while such a phenomenology may be nonfoundational and hermeneutic in at least Rorty's sense, it also may give the appearance of having a vestige of the previous past.

Phenomenology, even if nonfoundationalist, remains *structural*. But its structuralism is of an odd sort. Within its chosen field of investigations—contexts of possibility which constitute possible experiences of the language-perception type—its still essentially investigative thrust is one which discovers (vestiges of truth seeking) a multiplicity of structures. I return once more to the notion of multistability to make the point.

Structures discovered are not all of one type. In the often used examples of visual multistability in the preceding chapters, I would contend that the structure of possibility is linear and arbitrary (contrary to Merleau-Ponty). But such a possibility structure in no way exhausts the possibilities of others. Were one to move from the abstract, two-dimensional drawing examples, to concrete, three-dimensional objects in the normal earthbound context, one might discover a structure of *graded* possibilities.

Vary the Washington Monument: Its current stability is upright (see fig. 4.1).

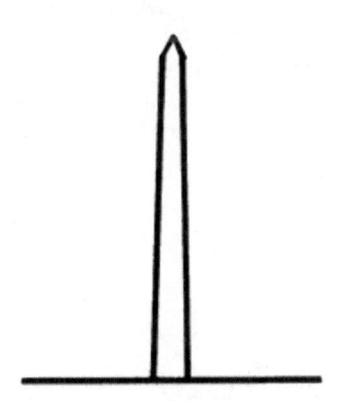

Figure 4.1. Washington Monument. Drawing created by Don Ihde.

This actual stability, without changing its architecture, would be even more stable, hence a graded possibility, were it to lie on its side on the level ground (see fig. 4.2).

Figure 4.2. Washington Monument on Its side. Drawing created by Don Ihde.

And under some temporary conditions it *might* even be (barely) stable upside down and perfectly balanced (see fig. 4.3).

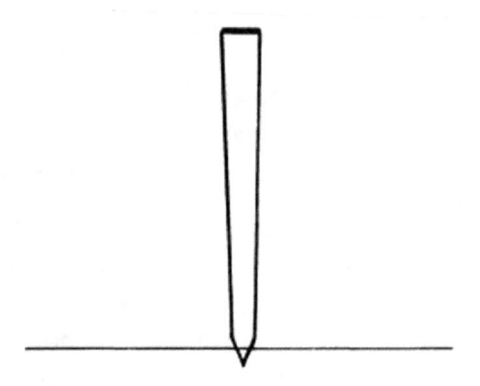

Figure 4.3. Washinton Monument Upside Down. Drawing created by Don Ihde.

But this last possibility is clearly gradedly weaker than either of the first two, while other possible positions are so weak as to be impossible (without changing the structure of the Monument itself) (see fig. 4.4).

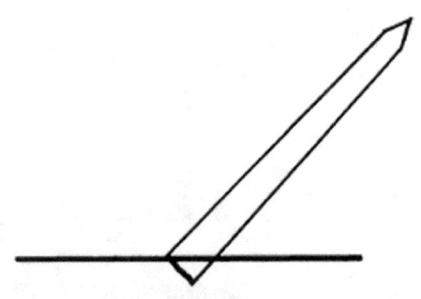

Figure 4.4. Washington Monument Diagonal. Drawing created by Don Ihde.

Still other types of structural stability could be hierarchical, serial, independent/dependent, etc. If this is vestigial "foundationalism" it is both oddly so, since the investigation and horizontalization of the field of structure is neither selective (all are context relative) nor reductive (there is no "best" or "only" structure). But it does come short of the one aspect of Rorty's conversations which place him much closer to the newer French versions of Continental philosophy than the older phenomenological ones.

By removing both truth seeking and referentiality entirely from edifying philosophy, Rorty joins the ranks of the post-structuralists and deconstructionists who have, while genuinely creating a new type of historical and cultural "science," also simply sidestepped the possibilities of what I prefer to call a *noematic* science. (The natural sciences, interestingly enough, come closer to this sense of phenomenological praxis—of the investigation of possible structures—than the previous human sciences.) This is in no way to deny what Kuhn and the new philosophy of science has discovered; that the practice of the sciences falls under essentially hermeneutic interpretations, that science itself includes an inevitably hermeneutic dimension, and so forth. But the need for the "outward" look, the noematic reference, is implicitly the retention of the language-*perception* pairing found within phenomenology.

Phenomenology insofar as it is essentially hermeneutic is edifying in Rorty's sense. That is, it is nonfoundational in its newer and post-Husserlian forms. But it is also edifying in a stronger sense for it is not without "edifice," *structure*. Finally, while phenomenology does have its vocabulary, it also retains its perceptions.

References

Aylesworth, Gary Eben. 1986. "From Grounds to Play: A Comparative Analysis of Wittgenstein and Heidegger." Doctoral thesis, State University of New York at Stony Brook.

Ihde, Don. 1982. "Phenomenology and Deconstructive Strategy." *Semiotica* 41:5–24.

———. 1983. *Existential Technics*. Albany: State University of New York Press.

Rorty, Richard. 1979. *Philosophy and the Mirror of Nature*. Princeton, NJ: Princeton University Press.

Chapter 5

What Is Postphenomenology?

This book [*Postphenomenology and Technoscience: The Peking University Lectures*] provides, in four chapters, a perspective on a very contemporary development stemming from my background in phenomenology and hermeneutics as directed toward science and technology. I have coined a special terminology, reflected in the title, *postphenomenology and technoscience*. And while a *post*-phenomenology clearly owes its roots to phenomenology, it is a deliberate adaptation or change in phenomenology that reflects historical changes in the twenty-first century. And, in parallel fashion, *techno*science also reflects historical changes that respond to contemporary science and technology studies. It is my deep conviction that the twentieth century marked radical changes with respect to philosophies, the sciences, and technologies. And this is clearly the case regarding the *interpretations* of these three phenomena. I illustrate this by referring to what has been called, in Anglophone countries, the "science wars." The American version, some would hold, began with the 1996 publication of the article "Transgressing the Boundaries: Towards a Transformative Hermeneutics of Quantum Gravity" in *Social Text*. The author, Alan Sokal, was a relatively unknown physicist, and the article was a deliberate hoax designed to show the ignorance of literary theorists and humanities academics. *Social Text* is a radical literary theory journal, and its board of editors was fooled into accepting and publishing the spoof. All of this escalated within the academy, in newspapers, and on the Internet. Stated broadly, the "wars" were about whether or not science is a universally valid, privileged mode of knowledge, culture and value free—this was the stance of the "science warriors." The literati, who were the brunt of the hoax and attack, were thought to have attacked science; they were claimed to be relativists, denying universal and absolute truth, for whom all modes of knowledge are simply subjective (the usual targets here were deconstructionists, feminists, and "social constructionists"). The vast popular discussion, of course, made Sokal "rich and famous," and the aftermath included a whole series of books, articles, and television shows.

This was the American version, but as I had already pointed out in my *Technology and the Lifeworld* (1990), a British version had pre-dated this set of battles. In 1987, *Nature*—surely one of the top science magazines—had included an opinion piece, again by two physicists, T. Theocharis and M. Psimopoulos, "Where Science Has Gone Wrong." Their thesis was that the decrease in support and funding, particularly in the Thatcher era, was due to the relativism of *philosophers of science*, and printed mug shots of Paul Feyerabend, Imre Lakatos, Karl Popper, and Thomas Kuhn headed the article. The objection was that these philosophers of science had undermined the belief in the universality, absoluteness, and value-and-culture free knowledge produced by science. And while this debate did not become as popularized as the later Sokal affair, it did continue on the pages of *Nature* for more than a year, until it was cut off by the editors.

In both of these cases, the "war" was over whether science is to be understood as acultural, ahistorical, universal, and absolute in its knowledge, or whether it is embedded in human history and culture and inclusive of the usual human fallibilities of other practices. Permit me now to *reframe* these incidents differently: One can also see these "wars" as *wars of interpretation*. That is, the context in which these events and controversies take shape includes such questions as: *What* is the most adequate interpretation of science? *Who* has the right to make such interpretations? *From what* perspectives do such interpretations take place? My two examples were of physicists playing the role of science expert interpreters. But what of others? The philosophers, historians, and social scientists? In short, I am suggesting that we "hermeneuticize" these phenomena.

I limit myself to the twentieth- and twenty-first-century context I have set here, roughly the period 1900–2006. With respect to early-twentieth-century interpretations of science, most of the best-known interpreters were philosophers who were trained in or practiced as scientists, including Pierre Duhem, Ernst Mach, and Henri Poincare, in the first decade of the century. These thinkers were trained in mathematics and/or physics. In short, this was a kind of *insider*, or, as it is now known, *internalist* interpretation. Similarly, when historians began to be interested in science interpretation, they were sometimes also trained in the sciences, or they looked at the historiography of science as a kind of heroic biography—great men had great ideas and produced great theories. This kind of history is still favored by many scientists as a preferred history of science.

We can now retrospectively recognize the emergence of both *positivist* and *phenomenological* variants on the philosophies of science. The famous Vienna Circle was formed on the one side, and the Gottingen School, including Husserl, on the other—and recall that Husserl's cognate disciplines remained logic and mathematics! To generalize, virtually all early *internalist* interpreters tended to model their interpretations upon science—and early phenomenology under Husserl conceived of itself as rigorous science. Phenomenology, from its beginnings, was one of the players in the early science interpretation wars.

All of this began to change by mid-century. By the onset of World War II, Husserl had died, and many of the positivists had emigrated to America, where they simply took over most American philosophies of science. Indeed, many emigrant philosophers believed that philosophy itself was equivalent to the philosophy of science. This stance, however, was not unchallenged and I trace its history in briefest form:

- The 1930s through the 1950s remained strongly held by logical positivism or logical empiricism with respect to the philosophy of science. The image of science was that of a sort of "theory-producing machine," which was verified through logical coherence and experiment.

- By the 1950s to the 1960s, a new antipositivist set of the philosophies of science emerged—Thomas Kuhn and kin, those mentioned in the *Nature* controversy—which added both histories and revolutions to the notion of science practice. Antipositivism remained theory centered but added discontinuous phenomena to early logicism. Historical particularity becomes part of interpreted science, "paradigm shifts." This image of science began to be enriched by historical sensitivity. Rather than a linear, cumulative historical trajectory, the antipositivists projected a narrative filled with "paradigm shifts" and punctuated discontinuities.

- The 1970s saw the emergence of new sociologies of scientific knowledge—"social constructionism" and "actor network theory" examined science in its social, political, and constructive dimensions. Science is seen as a particular social practice. Its results were viewed as negotiated and constructed.

- In the 1980s, new philosophies of technology (post-Heideggerian, post-Ellul, post-Marxian) introduced the recognition that science itself is also *technologically embodied*. Without instruments and laboratories, there was no science.

- In the late 1980s and 1990s, feminist philosophies began to locate patriarchal biases in science practice, which in some cases led to new understandings of reproductive strategies in evolution. Science was seen as frequently gendered in cultural practice.

The combined result, decried by the reactions of the science warriors, was that science was now seen as fully acculturated, historical, contingent, fallible, and social, and whatever its results, its knowledge is produced out of practices. I contend that by the end of the twentieth century, even those belonging to the analytic versions of the philosophy of science could be seen to have made concessions. For example, Ernan McMullen of the dominantly analytic philosophy of science department at Notre Dame University edited a book called *The Social Dimensions of Science* (1992), clearly acknowledging the now-richer image of science than that of a "theory-producing machine." And Larry Laudan, in his *Science and Relativism* (1990), which is a debate among varieties of analytic philosophies of science, proclaims that all are now "fallibilists."

I take it that this was the consensus at the end of the twentieth century, brought about by a now-widened, more diverse set of interpreters. However, the now enlarged field of interpreters also may be seen retrospectively as a response to the obvious massive changes to both science and technology in the twentieth century. For instance, from 1900 to 2006, one can see that big science, corporate science, and global science are the order of the day. From the Manhattan Project to the Human Genome Project, from physics to biology, there is big science. And the same radical change in technologies should be even more obvious: in 1900, there were no airplanes, no nuclear energy, no computers or Internet, and so on, whereas today these constitute the texture of our very lifeworld. And now my special move: I want to place philosophy, particularly phenomenology, precisely into this scene and interpret it, judge it, through a series of changed interpretations parallel to those used to interpret science and technology. What is philosophy, phenomenology, from a contemporary perspective? Philosophy, too, I hold, changes, or must change with its historical context. This is what produces my attempt to modify classical phenomenology into a contemporary *postphenomenology*. So it is now time to briefly look at phenomenological philosophy, roughly in the same 1900–2006 period relevant here. I do this by first looking at the interrelationship between phenomenology and pragmatism.

First Step: Pragmatism and Phenomenology

Phenomenology in Europe and pragmatism in America were historically simultaneously born. Both were new, radical philosophies that placed *experience* in a central role for analysis. Pragmatism was first called so by William James (1898), who credited it to Charles Sanders Peirce; William James also was an early major influence on Husserl, but pragmatism was brought to prominence primarily by John Dewey. Note that Dewey and Husserl were both born in 1859, and although Dewey lived longer than Husserl, their philosophical developments were chronologically parallel. But also note that their birth year was also the same as the publication of Darwin's *Origin of Species*. Or, since 2005 was the centennial of Einstein's golden year, 1905, if we also look at Dewey in 1905, we find him at Columbia University, already famous in the philosophy of education after founding his earlier experimental or laboratory school at the University of Chicago. And, if we look at Husserl in 1905, we find him giving his internal time lectures.

In terms of the historical philosophical context at the turn of the century, there were some similarities but also nuanced differences between the pragmatists and Husserl's phenomenology. This can be subtly illustrated in the term *pragmatism* itself. Dewey, in his "The Development of American Pragmatism," says, "The term 'pragmatic,' contrary to the opinion of those who regard pragmatism as an exclusively American conception, was suggested to [Peirce] by the study of Kant . . . in the *Metaphysics of Morals* Kant established a distinction between *pragmatic* and the *practical*. [Practical] applies to the moral laws which Kant regards as *a priori* . . . whereas [pragmatic] applies to the rules of art and technique which are *based on experience and are applicable to experience*" (Dewey 1977, 42; emphasis added). Now, as we know, Descartes and Kant also play major roles in Husserl's development of phenomenology—but the roles they play are those of an *epistemological* Descartes and Kant, whereas it is the *moral* but also a "praxical" Kant who is used by Peirce! The pragmatic emphasis is on *practice*, not *representation*. This move to praxis and away from representation later repeats itself in virtually all the late-twentieth-century styles of science interpretation.

This different take on Kant is subtle and nuanced, but I want to make a very bold extrapolation from this difference: By using the epistemological Descartes and Kant, Husserl necessarily had to also use the vocabulary of early modern "subject/object," "internal/external," "body/mind," as well as "ego," "consciousness," and the like. And while it is clear that his attempt was to *invert* these usages through the use of his various *reductions*, this vocabulary remained embedded in early phenomenology. This attempt to overcome early modern epistemology, while using its terminology, I contend, doomed classical phenomenology to be understood and interpreted as a "subjective" style of philosophy. The pragmatists, by beginning with the vocabulary of practices instead of representations, avoided this problem. Listen to a contemporary pragmatist echoing this idea: Richard Rorty says, "The pragmatists tell us it is the vocabulary of practice rather than theory, of action rather than contemplation, in which one can say something about truth. . . . My first characterization of pragmatism is that it is simply anti-essentialism applied to notions like 'truth,' 'knowledge,' 'language,' 'morality,' and similar objects of philosophical theorizing. . . . So, pragmatists see the Platonic tradition as having outlived its usefulness. This does not mean that they have a new, non-Platonic set of answers to Platonic questions to offer, but rather they do not think we should ask those questions anymore" (Rorty 1982, 197).

Returning to Dewey, his early writings contain many essays on the new science, *psychology*. This psychology—although for Dewey the outdated philosopher to be transcended was more Locke than Descartes—proposed to analyze *consciousness*. And whereas Husserl, too, had a problem with psychologism, Dewey again seems to cut to the core more quickly. For him, "consciousness" in psychology is an *abstraction*, whereas experience is broader and necessarily related to other dimensions "if the individual of whom psychology treats be, after all, a social individual, any absolute setting off and apart of a sphere of consciousness as, even for scientific purposes, self-sufficient, is *condemned in advance*" (Dewey 1977, 161–62; emphasis added). While Husserl's inversion of Descartes includes "all subjectivity is intersubjectivity," Husserl arrives late at such a recognition. I cannot go much farther here, but one clue to pragmatism's quicker take on the problems of early modern epistemology also may lie in its recognition that there is a biological, evolutionary dimension to "psychology." Put simply, Dewey's frequent model or metaphor for his version of transformational practice is that of an organism/environment model rather than a subject/object model. Again, turning to Dewey's early writings, "In the orthodox view, experience is regarded primarily as a knowledge-affair [Locke/Descartes]. But to eyes not looking through ancient spectacles, it assuredly appears as an affair of the intercourse of a living being with its physical and social environment" (Dewey 1977, 61).

This living being/environment model, for Dewey, is also "experimental," and thus less past or present directed than future directed. Experience in its vital form is experimental, an effort to change the given; it is characterized by projection, by reaching forward into the unknown; connection with a future is its salient trait" (Dewey 1977, 61). (Interestingly, this future emphasis seems closer to Heidegger than to Husserl.) Dewey sees this model as "biological" in some sense, and he imputes this both to one phase of William James's version of psychology, but also to Darwin, whose notion of change-through-time also outlines the points just made. Once again, my contention is that this version of experience short-circuits the "subject/object" detour derived from Descartes—or, in Dewey's case, Locke—and points much more directly to something like a *lifeworld analysis*.

Now, admittedly, I have here the advantage of retrospective vision; I am looking at Dewey and Husserl, pragmatism and phenomenology, from a full century later perspective. But it remains the case that there were resources then contemporarily available from pragmatism, which had Husserl used them would have yielded a *nonsubjectivistic and interelational phenomenology* along the lines I am now calling *postphenomenology*. This is why I have here paralleled Husserl and Dewey, who were exact contemporaries. This grafting of pragmatism to phenomenology constitutes a first step in a postphenomenological trajectory.

Second Step: Phenomenology and Pragmatism

In my first step, I suggested that the deconstruction of early modern epistemology made in pragmatism could have enriched the beginnings of phenomenology by avoiding the problems of subjectivism and idealism with which early phenomenology was cast. My second step reverses the process, and I now suggest that phenomenology historically developed a style of rigorous analysis of experience that was potentially *experimental* and thus relevant to pragmatism. Dewey's emphasis on his experience-based philosophy was "experimental," or sometimes called "instrumental," but I contend that Husserl's phenomenology contained methods that, had these been adapted in pragmatism, would have enriched its analysis of the experimental. In this case, however, rather than return to Husserlian observations from his texts, I shall instead take these for granted and draw three elements from phenomenology to show how such a rigorous analysis of the experiential takes shape. These include: *variational theory, embodiment*, and the notion of *lifeworld*. Phenomenologists will recognize that all three may be found in Husserl, although I would claim that embodiment was later highly enriched by Merleau-Ponty, and that what could be called the cultural-historical dimensions of the lifeworld were correspondingly enriched by Heidegger. Each of these notions derives from classical phenomenology, but each now takes their shape and role in a contemporary postphenomenology.

I begin with *variational theory*: In Husserl's earlier use, variations (originally derived from mathematical variational theory) were needed to determine *essential structures*, or "essences." Variations could be used to determine what was variant and what invariant. I also have found this technique invaluable in any phenomenological analysis—but as I used this technique, I discovered something other than Husserlian "essences" as results. What emerged or "showed itself" was the complicated structure of *multistability*. My first systematic demonstration of this phenomenon occurred in *Experimental Phenomenology* (1977). Using so-called visual illusions, I tried to show how the phenomenological notion of variation yielded both deeper and more rigorous analyses of such illusions than mere empirical or psychological methods. To demonstrate this analysis, I draw from three example sets from those studies:

In the first example, stage/pyramid/robot, this configuration, an abstract drawing, can be seen as a stage setting (fig. 5.1).

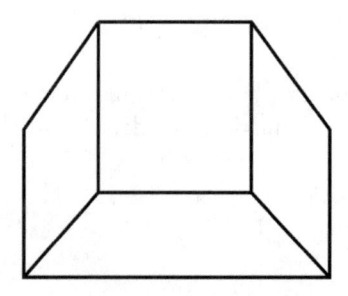

Figure 5.1. Multistable Figure A. Drawing created by Don Ihde.

The plane surface at the bottom of the drawing is the stage, while the other surfaces are the backdrops. Thus an apparent three-dimensionality appears—but it also implies a perspective from which this three-dimensionality takes its shape. The POV, or "point of view," is a sort of balcony position from which the viewer looks slightly downward at the stage setting. Here already, then, *embodiment*, or perspectival perception, is implied. But this is only one variation—the *same* configuration could be seen quite differently. Perhaps it is a Mayan pyramid in Central America! In this case the plane surfaces change appearances: the center, upper surface is now the platform on the top of the pyramid, and the other surfaces are the downward sloping sides. In this appearance, the three-dimensionality is radically reconfigured but remains three-dimensional. And the POV, or perspective, also remains implied—as if we are in a helicopter viewing the pyramid from above. Note too that these two appearances are discrete and different—they are *alternations* which cannot be combined; they are distinct variations. As an aside to empirical studies, such three-dimensional reversals are well known in psychology—particularly in gestalt psychology, there called "gestalt switches." And while historically the early gestaltists were in fact students of Husserl, we have not left "psychology" quite yet. Now, the first phenomenologically deeper move: I suggest that there is another possible stability here. My story is that this configuration also may be seen as a "headless robot." In this case what was previously the platform of the pyramid now becomes the robot's body. The bottom line is the earth on which the robot is walking, and the other lines are its arms and legs, and—because it has no head—it uses crutches to navigate! In this configuration, three-dimensionality is lost, and the figure is simply two-dimensional. But take careful note: in the two-dimensional appearance, the implied POV, or embodied position, *also changes. Now it is directly before the robot, who is advancing toward the viewer!* This is all fully phenomenological: variant perceptual profiles, examined through variations/implied perceptual-bodily positions, which correlate to and change with the appearances/and, now, more than occur in mere empirical studies.

My next set of illustrations comes from the famous Necker Cube series (fig. 5.2).

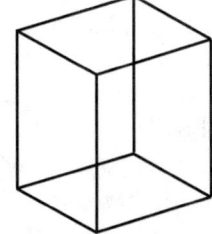

Figure 5.2. Multistable Figure B. Drawing created by Don Ihde.

When I was writing my book *Experimental Phenomenology*, I read over 1,000 pages of studies on the Necker Cube phenomenon, all of which recognized the three-dimensional reversal, and a few of which recognized a two-dimensional variant (but usually associated with a "fatigued subject" rather than a noematic possibility within the configuration). Quickly, now, one can easily see that the Necker Cube may be seen as three-dimensional, with a "tilt" switch. Note that there is also a small but detectable switch in the POV, or perspective position, in the switch. To make this into a two-dimensional variation, a new story may be told: I tell you that this is not a cube at all but an insect in a hexagonal hole. The limits of the cube are now the outline of the hole; the central surface is the body of the insect; and the other lines are its legs. Now the figure becomes two-dimensional— and again the POV is directly correlated to the insect. You can easily see that—so far—the Necker Cube has the *same structural set of possibilities as the previous example*, and that the shifts of position, implied embodiment, are all parallel. But since the empirical literature sometimes, though rarely, recognized the two 3-D and one 2-D variations, phenomenology has not yet gotten deeper than gestalt psychology—but it can. Return to the configuration with a new story: what was previously the insect's body now becomes the forward-facing facet of an oddly cut gem. The various surfaces around this central facet are the other facets of this gem—and once you see this, you can immediately tell that this is again three-dimensional in appearance, but in a totally different way than previously as a cube. And, now, if you are learning fast, you can anticipate that a *reversal of this three-dimensionality is also possible*. One is looking from "inside" or from the bottom of the gem and the once-forward facet is not the distant facet. Add quickly, and we have "constituted" *five* variations so far, not three, as in gestalt psychology, and thus once again phenomenological variations go farther than empirical psychology.

My third example set is slightly different than the previous two. In both the stage/pyramid series and the Necker Cube series, the variants were all discrete, distinct, and alternations were not commensurable with each other. Each had multistabilities but discrete stabilities. In this example there is a continuity phenomenon that nevertheless retains its own kind of multistability. This example is the famous Hering Illusion. Here, as one looks at the configuration, the claim is made that the two horizontal lines "appear" to be bent, but in "reality" they are straight. (This appearance/ reality distinction, presupposed from modernist metaphysics, is what makes this an "illusion." Phenomenologically, of course, reductions eliminate the appearance/reality distinction in favor of what "shows itself.")

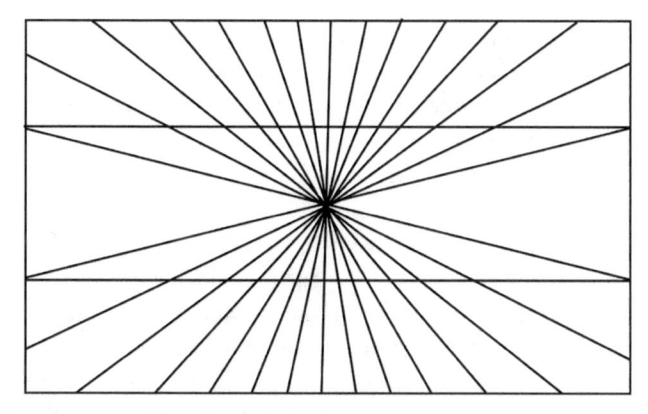

Figure 5.3. Multistable Figure C. Drawing created by Don Ihde.

Now my phenomenological deconstruction of this "illusion" is attainable as follows: focus upon the convergence of lines at the center of the drawing; now "push" this point into distant infinity. As you do this, the horizontal lines *straighten*. Now reverse the process and bring the point that is at infinity back toward your viewing position. You will see the horizontal lines recurve and then straighten out with the three-dimensional reversal. So here again we have multistability, and, as in the other cases, it is related to variations upon two- and three-dimensionality—but also to the context in which straight and curved show a continuous structure. Empirical psychology simply assumed a sedimented and nondepth view, which through deliberate variation shows change. Phenomenologically, perception is not passive but active; holistically, it is bodily interactive with an environment, but while this agrees with both pragmatism and phenomenology, it is the phenomenologically derived *variation* that provides the rigorous demonstration.

We are now partway with step two, the phenomenological enrichment of pragmatism into a postphenomenology. And while I have just made variational theory the central method to give rigor to experiential analysis, the implicit role of embodiment also came into play. Active perceptual engagement, implied in all of the example sets, reveals the situated and perspectival nature of bodily perception—again, an important point repeated by Husserl in his classical analyses (profiles, latent and manifest presences, sometimes even applied to a solid cube in his examples).

At this point I want to make a large leap to an example set now related to *technologies*. While the use of visual "illusions" has the advantage of initial clarity and ease to demonstrate multistability as a phenomenological result of variational analyses, these illustrations also have the disadvantage of being all too simple and all too abstract. This is particularly the case with respect to the weak sense of embodiment in the illusions set. My POV, or perspective, is clear but weak in the sense that I am in a mere "observer" position vis-á-vis the examples. So my next example set will draw from a very ancient, very simple, and very multicultural set of technologies: *archery* (bows and arrows). Although I have researched, and continue to research, the history of archery, I do not believe anyone knows who or where it was first invented. I did meet someone in XiAn who claimed that the Chinese invented archery—in the history of technology, the usual claim is "the Chinese did it first"—but in this case they did not. Some arrowheads date back to at least 20,000 BP; there is an embedded arrowhead in a skeletal pelvis dated 13,000 BP. And, in this case, some European arrowheads date back to 11,000 BP. Then there is Otzi, the freeze-dried mummy found in the Italian Alps in 1991, carbon dated back to 5,300 years ago, who had a full archery set with him, two millennia earlier than the 3,000-year-old Chinese treatises on archery. (There is evidence, however, that the Chinese did first invent the *crossbow*, one of which is displayed in XiAn with the chariots recovered there.) In any case, except for Australia, where boomerangs are used, and parts of equatorial jungles, where blowguns are used—rare cultures in which archery never occurred—virtually all ancient cultures had bows and arrows.

My use here, however, is to show how this practice is also *multistable* in precisely its phenomenological sense developed in the earlier examples. Once again I look for variations, embodiment, and now, more fully, lifeworld dimensions. In an abstract sense, all archery is the "same" technology in which a projectile (arrow) is propelled by the tensile force of a bow and bowstring. But as we shall see, radically different practices fit differently into various contexts:

The first example is the English longbow (fig. 5.4).

Figure 5.4. English Longbow. Drawing created by Max Liboiron.

One famous battle often referred to in European history is that of the English versus the French at Agincourt. This battle was one not only of nationalities but of technologies—the French preferred the crossbow, the English the longbow. Both were powerful weapons, but while the crossbow was somewhat more powerful, it also was slow compared to the rapid fire capacity of the longbow. At this battle, 6,000 bowmen withstood and prevailed over 30,000 infantry and knights. Consider now the material technology, the bodily technique, and the social practice of the longbowmen: the bow was made of yew, usually about six feet or two meters long. It was held by bowmen in a standing position, and the bow was held out in front in a stable position. The bowstring was pulled back toward the eye of the soldier, with four fingers on it, and released when the aim was proper. Arrows were available either in a quiver or stuck in the ground, and firing was fast.

The second example is mounted archery, used by Mongolian horsemen and in the early medieval invasions of Eastern Europe (see fig. 5.5).

Figure 5.5. Mongolian Horse Bow. Drawing created by Max Liboiron.

The horsemen used archery while mounted on speeding horses. While one could say that mounted archers used the "same" bow and arrow technology as weaponry during the Mongol invasions, in another sense there were radical, alternative aspects to horse-mounted archery. First, the bow, was short—rarely more than a meter or a little more—made of composite materials (bone, wood, skin and glue), and deeply *recurved*. The power of the bow was similar to that of the long bow but had less distance-gaining capacity. The bodily technique also was radically different. Used while at a gallop, the archer held the bowstring near his eye and pushed the bow outward for rapid fire. (Although not recurved, American Indians used a similar technique for buffalo hunting.)

The third example is "Artillery archery," what I shall call the ancient Chinese archery that utilized the most powerful of all premodern bows known (see fig. 5.6).

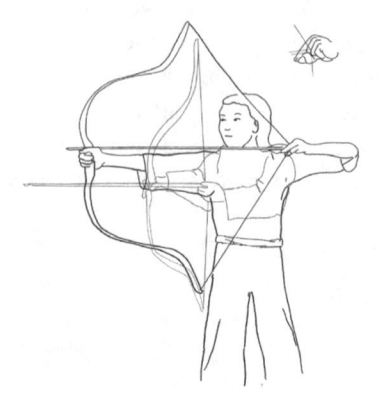

Figure 5.6. Chinese "Artillary" Bow (with Thumb Ring). Drawing Created by Max Liboiron.

The pull needed for these long and partially recurved bows was in the 140-pound range. Here the technique called for a *simultaneous push and pull* to launch the arrow, and a unique use of the thumb, with a thumb ring, was required for the bowstring. (I had learned of this technique before actually visiting XiAn in 2004, but during my visit I was delighted to see the terra-cotta archers precisely positioned for this technique!) So what we see again is another stability in which the actual materiality of the bow, the bodily technique of use, and the cultural-historical role this technology plays as a variant.

I am not claiming here to have exhausted the variations, but these three are enough to show that the phenomenological variations that now include considerations of the materiality of the technologies, the bodily techniques of use, and the cultural context of the practice are all taken into account and demonstrate again the importance of variational theory with its outcome in multistability, the role of embodiment, now in trained practice, and the appearance of differently structured lifeworlds relative to historical cultures and environments.

I have now also illustrated the pragmatism to phenomenology and the phenomenology to pragmatism moves needed to outline an initial postphenomenology. The enrichment of pragmatism includes its recognition that "consciousness" is an abstraction, that experience in its deeper and broader sense entails its embeddedness in both the physical or material world and its cultural-social dimensions. Rather than a philosophy of consciousness, pragmatism views experience in a more organism/environment model. The reverse enrichment from phenomenology includes its more rigorous style of analysis that develops variational theory, recognizes the role of embodiment, and situates this in a lifeworld particular to different epochs and locations. There remains one more step to make what I am now calling postphenomenology fully contemporary. That is the inclusion of a "science"—better, "technoscience studies" approach to our contemporary lifeworld.

Third Step: "The Empirical Turn"

I began this chapter with a glance at the "science wars" that grew out of issues of interpretation concerning science, technology, and philosophy for purposes here. My contention is that science, technology, and *philosophy* have all undergone major changes through the twentieth into the twenty-first centuries. And while the next chapter [of *Postphenomenology and Technoscience*] will focus on those changes, before concluding my outline of postphenomenology, I turn to one more episode in its construction. In this case we have to move beyond both classical pragmatism and classical phenomenology and into the realm of the philosophy of technology. Neither Dewey nor Husserl made dealings with material technologies thematic. Dewey recognized that psychological experience was a mere abstraction unless it took into account both the physical world and the social world. And while he did parallel Heidegger with respect to the insight that technologies *precede* science, and that science cannot exist without technologies, he did not engage in analyses that would specifically highlight our experience *of* technologies. In Husserl's case, there are few references to technologies at all. The closest he comes—as I have held elsewhere—is in his recognition of measurement practices lying at the base of the origin of geometry, and his recognition that writing raises consciousness to a higher level.

Martin Heidegger was the exception in classical phenomenology, since by wide agreement he may be considered a major thinker at the origins of the late modern philosophy of technology. I also shall return to Heidegger in the next chapter, but in this setting I will "leapfrog" his work in order to outline the third step constituting postphenomenology. That step is what Dutch philosophers of technology have called "the empirical turn," a phrase that has caught on and is now widely used to describe in particular the very contemporary philosophy of technology.

Here is the context: The Netherlands has had a strong tradition in the philosophy of technology, dating going back to the early twentieth century, and one of its main centers today is at the University of Twente. Hans Achterhuis, himself a leading philosopher of technology, collaborated with his colleagues and published in 1992 the book, *De Matt van de Techniek* [The Measure or Metier of Technics]. This book could be thought of as dealing with the early twentieth-century foundations of the philosophy of technology. It dealt with the first twentieth-century founders of the philosophy of technology, including Martin Heidegger and Jacques Ellul, but also Lewis Mumford and Hans Jonas. But in 1997, again with his colleagues, Achterhuis published a second book, *Van Stoommachine tot Cyborg: Denken over techniek in de neiuwe wereld*, literally translated as *From Steam Engine to Cyborg: Thinking Technology in the New World*. This book purports to show that a newer generation of philosophers of technology, six chosen from philosophy in America, has shifted the center of gravity by making "an empirical turn." I found this Dutch perspective an interesting one, and thus we had it translated into English (capably done by my colleague, Bob Crease) as *American Philosophy of Technology: The Empirical Turn* (2001).

There are three ways in which Achterhuis sees differences between the classical philosophy of technology and the contemporary philosophy of technology:

- Classical philosophers of technology tended to be concerned with technology overall and not specific technologies. "The classical philosophers of technology occupied themselves more with the historical and *transcendental* conditions that made modern technology possible than with the real changes accompanying the development of a technological culture" (Achterhuis 2001, 6–8; emphasis added).

- Classical philosophers of technology often displayed romantic or nostalgic tastes, thus displaying a dystopian cast to their interpretations of technology. "The issue [now] . . . is to understand this new cultural constellation, rather than to reject it nostalgically in demanding a return to some prior, seemingly more harmonious and idyllic relations assumed to be possible between nature and culture [as in the classical philosophy of technology]" (Achterhuis 2001, 6–8).

- Achterhuis notes that the new philosophers of technology took an empirical—or a concrete—turn described thus: "About two

decades ago, dissatisfaction with the existing, classical philosophical approach to technology among those who studied new
developments in technological culture as well as the design
stages of new technologies led to an empirical turn that might
roughly be characterized as constructivist. This empirical turn
was broader and more diverse than the one that had taken place
earlier in the philosophy of science, especially as inspired by
the work of Thomas Kuhn, but shared a number of common
features with it. First, this new generation of thinkers opened
the black box of technological developments. Instead of treating
technological artifacts as givens, they analyzed their concrete
development and formation, a process in which many different
actors become implicated. In place of describing technology
as autonomous, they brought to light the many social forces
that act upon it. Second, just as the earlier, Kuhn-inspired
philosophers of science refused to treat 'science' as monolithic,
but found that it needed to be broken up into many different
sciences, each of which needed to be independently analyzed,
so the new philosophers of technology found the same had to
be done with 'technology.' Third, just as the earlier philosophers
of science found that they had to speak of the co-evolution
of science and society, so the new, more empirically oriented
philosophers of technology began to speak of the co-evolution
of technology and society" (Achterhuis 2001, 6–8).

I accept this characterization of the contemporary set of philosophers of
technology included in Achterhuis's book. Furthermore, this description is
what I am calling the third step toward a postphenomenology. It is the step
away from generalizations about *technology uberhaupt* and a step into the
examination of *technologies in their particularities.* It is the step away from
a high altitude or transcendental perspective and an appreciation of the
multidimensionality of technologies as *material cultures* within a *lifeworld.*
And it is a step into the style of much "science studies," which deals with
case studies.

As Achterhuis correctly recognizes, such a step is not one that occurs
in isolation; rather, it reflects precisely the broad front common to most new
interpreters of science and technology. The new philosophies of science, the
new sociologies of science and feminism, and now the new philosophies of
technology all, to some degree, and each in their way, become more concrete
in their examinations of what I call *technoscience.*

If this, then, is the contemporary philosophy of technology, then I want to make one final observation about this position compared to both the classical beginnings of pragmatism and phenomenology. As noted earlier, neither Dewey nor Husserl made technologies as such thematic to their philosophies. In Dewey's case, there remained a broad, modernist concern with the natural world and the social world. The experiencer—the human—related to both the physical and the social was thought of as an organism within an environment, in Husserl's case, the "World," or his equivalent of an environment, was also made up of things and of the problematic presence of others, as in the *Cartesian Meditations* and, later, with the historical-cultural-"praxical" world of the *Crisis*. In neither were relations with technologies as such made thematic or specific. With the arrival of the philosophy of technology, which in its dominant form arose from the *praxis traditions* of philosophy—pragmatism, phenomenology, Marxism—the thematization of human experience in relation to technologies produced a changed philosophical landscape.

Such a thematization, however, includes perhaps the farthest-reaching modification to classical phenomenology. In both pragmatism and phenomenology, one can discern what could be called an *interrelational ontology*. By this I mean that the human experiencer is to be found ontologically related to an environment or a world, but the interrelation is such that both are transformed within this relationality. In the Husserlian context, this is, of course, *intentionality*. In the context of his *Ideas* and *Cartesian Meditations* this is the famous "consciousness *of* ___," or all consciousness is consciousness of "something." I contend that the inclusion of technologies introduces something quite different into this relationality. Technologies can be the means by which "consciousness itself" is *mediated*. Technologies may occupy the "of" and not just be some object domain. This theme recurs later in this book.

What Is Postphenomenology?

Postphenomenology is a modified, hybrid phenomenology. On the one side, it recognizes the role of pragmatism in the overcoming of early modern epistemology and metaphysics. It sees in classical pragmatism a way to avoid the problems and misunderstandings of phenomenology as a subjectivist philosophy, sometimes taken as antiscientific, locked into idealism or solipsism. Pragmatism has never been thought of this way, and I regard this as a positive feature. On the other side, it sees in the history of phenomenology a development of a rigorous style of analysis through the use of variational theory, the deeper phenomenological understanding of embodiment and human active bodily perception, and a dynamic understanding of a lifeworld as a fruitful enrichment of pragmatism. And, finally, with the emergence of the philosophy of technology, it finds a way to probe and analyze the role of technologies in social, personal, and cultural life that it undertakes by concrete—empirical—studies of technologies in the plural. This, then, is a minimal outline of what constitutes *postphenomenology*.

References

Achterhuis, Hans. 2001. *American Philosophy of Technology: The Empirical Turn.* Bloomington: Indiana University Press.

Dewey, John. 1977. *The Philosophy of John Dewey.* Edited by John J. McDermott. New York: Putnam.

Rorty, Richard. 1982. *Consequences of Pragmatism.* Princeton, NJ: Princeton University Press.

Part 2

The Phenomenology of Technology

Chapter 6

Human-Technology Relations

The task of a phenomenology of human-technology relations is to discover the various structural features of those ambiguous relations. In taking up this task, I shall begin with a focus upon experientially recognizable features that are centered upon the ways we are bodily engaged with technologies. The beginning will be within the various ways in which I-as-body interact with my environment by means of technologies.

Technics Embodied

If much of early modern science gained its new vision of the world through optical technologies, the process of embodiment itself is both much older and more pervasive. To embody one's praxis *through* technologies is ultimately an *existential* relation with the world. It is something humans have always—since they left the naked perceptions of the Garden—done.

I have previously and in a more suggestive fashion already noted some features of the visual embodiment of optical technologies. Vision is technologically transformed through such optics. But while the fact *that* optics transform vision may be clear, the variants and invariants of such a transformation are not yet precise. That becomes the task for a more rigorous and structural phenomenology of embodiment. I shall begin by drawing from some of the previous features mentioned in the preliminary phenomenology of visual technics.

Within the framework of phenomenological relativity, visual technics first may be located within the intentionality of seeing.

I see—through the optical artifact—the world

This seeing is, in however small a degree, at least minimally distinct from a direct or naked seeing.

I see—the world

I call this first set of existential technological relations with the world *embodiment relations*, because in this use context I take the technologies *into* my experiencing in a particular way by way of perceiving *through* such technologies and through the reflexive transformation of my perceptual and body sense.

In Galileo's use of the telescope, he embodies his seeing through the telescope thusly:

Galileo—telescope—Moon

Equivalently, the wearer of eyeglasses embodies eyeglass technology:

I—glasses—world

The technology is actually *between* the seer and the seen, in a *position of mediation*. But the referent of the seeing, that towards which sight is directed, is "on the other side" of the optics. One sees *through* the optics. This, however, is not enough to specify this relation as an embodiment one. This is because one first has to determine *where* and *how*, along what will be described as a continuum of relations, the technology is experienced.

There is an initial sense in which this positioning is doubly ambiguous. First, the technology must be *technically* capable of being seen through; it must be transparent. I shall use the term *technical* to refer to the physical characteristics of the technology. Such characteristics may be designed or they may be discovered. Here the disciplines that deal with such characteristics are informative, although indirectly so for the philosophical analysis per se. If the glass is not transparent enough, seeing-through is not possible. If it is transparent enough, approximating whatever "pure" transparency could be empirically attainable, then it becomes possible to embody the technology. This is a material condition for embodiment.

Embodying as an activity, too, has an initial ambiguity. It must be learned or, in phenomenological terms, constituted. If the technology is good, this is usually easy. The very first time I put on my glasses, I see the now-corrected world. The adjustments I have to make are not usually focal irritations but fringe ones (such as the adjustment to backglare and the slight changes in spatial motility). But once learned, the embodiment relation can be more precisely described as one in which the technology becomes maximally "transparent." It is, as it were, taken into my own perceptual-bodily self experience thus:

(I-glasses)-world

My glasses become part of the way I ordinarily experience my surroundings; they "withdraw" and are barely noticed, if at all. I have then actively embodied the technics of vision. Technics is the symbiosis of artifact and user within a human action.

Embodiment relations, however, are not at all restricted to visual relations. They may occur for any sensory or microperceptual dimension. A hearing aid does this for hearing, and the blind man's cane for tactile motility. Note that in these corrective technologies *the same structural features of embodiment* obtain as with the visual example. Once learned, cane and hearing aid "withdraw" (if the technology is good—and here we have an experiential clue for the perfecting of technologies). I hear the world through the hearing aid and feel (and hear) it through the cane. The juncture (I-artifact)-world is through the technology and brought close by it.

Such relations *through* technologies are not limited to either simple or complex technologies. Glasses, insofar as they are engineered systems, are much simpler than hearing aids. More complex than either of these monosensory devices are those that entail whole-body motility. One such common technology is automobile driving. Although driving an automobile encompasses more than embodiment relations, its pleasurability is frequently that associated with embodiment relations.

One experiences the road and surroundings *through* driving the car, and motion is the focal activity. In a finely engineered sports car, for example, one has a more precise feeling of the road and of the traction upon it than in the older, softer-riding, large cars of the fifties. One embodies the car, too, in such activities as parallel parking: when well embodied, one feels rather than sees the distance between car and curb—one's bodily sense is "extended" to the parameters of the driver-car "body." And although these embodiment relations entail larger, more complex artifacts and entail a somewhat longer, more complex learning process, the bodily tacit knowledge that is acquired is perceptual-bodily.

Here is a first clue to the polymorphous sense of bodily extension. The experience of one's "body image" is not fixed but malleably extendable and/or reducible in terms of the material or technological mediations that may be embodied. I shall restrict the term embodiment, however, to those types of mediation that can be so experienced. The same dynamic polymorphousness can also be located in non-mediational or direct experience. Persons trained in the martial arts, such as karate, learn to feel the vectors and trajectories of the opponent's moves within the space of the combat. The near space around one's material body is charged.

Embodiment relations are a particular kind of use-context. They are technologically relative in a double sense. First, the technology must "fit" the use. Indeed, within the realm of embodiment relations one can develop a quite specific set of qualities for design relating to attaining the requisite technological "withdrawal." For example, in handling highly radioactive materials at a distance, the mechanical arms and hands which are designed to pick up and pour glass tubes inside the shielded enclosure have to "feed back" a delicate sense of touch to the operator. The closer to invisibility, transparency, and the extension of one's own bodily sense this technology allows, the better. Note that the design perfection is not one related to the machine alone but to the combination of machine and human. The machine is perfected along a bodily vector, molded to the perceptions and actions of humans.

And when such developments are most successful, there may arise a certain romanticizing of technology. In much anti-technological literature there are nostalgic calls for returns to simple tool technologies. In part, this may be because long-developed tools are excellent examples of bodily expressivity. They are both direct in actional terms and immediately experienced; but what is missed is that such embodiment relations may take any number of directions. Both the sports car driver within the constraints of the racing route and the bulldozer driver destroying a rainforest may have the satisfactions of powerful embodiment relations.

There is also a deeper desire which can arise from the experience of embodiment relations. It is the doubled desire that, on one side, is a wish for *total transparency*, total embodiment, for the technology to truly "become me." Were this possible, it would be equivalent to there being no technology, for total transparency would be my body and senses; I desire the face-to-face that I would experience without the technology. But that is only one side of the desire. The other side is the desire to have the power, the transformation that the technology makes available. Only by using the technology is my bodily power enhanced and magnified by speed, through distance, or by any of the other ways in which technologies change my capacities. These capacities are always *different* from my naked capacities. The desire is, at best, contradictory. I want the transformation that the technology allows, but I want it in such a way that I am basically unaware of its presence. I want it in such a way that it becomes me. Such a desire both secretly *rejects* what technologies are and overlooks the transformational effects which are necessarily tied to human-technology relations. This illusory desire belongs equally to pro- and anti-technology interpretations of technology.

The desire is the source of both utopian and dystopian dreams. The actual, or material, technology always carries with it only a partial or quasi-transparency, which is the price for the extension of magnification that technologies give. In extending bodily capacities, the technology also transforms them. In that sense, all technologies in use are non-neutral. They change the basic situation, however subtly, however minimally; but this is the other side of the desire. The desire is simultaneously a desire for a change in situation—to inhabit the earth, or even to go beyond the earth—while sometimes inconsistently and secretly wishing that this movement could be without the mediation of the technology.

The direction of desire opened by embodied technologies also has its positive and negative thrusts. Instrumentation in the knowledge activities, notably science, is the gradual extension of perception into new realms. The desire is to see, but seeing is seeing through instrumentation. Negatively, the desire for pure transparency is the wish to escape the limitations of the material technology. It is a platonism returned in a new form, the desire to escape the newly extended body of technological engagement. In the wish there remains the contradiction: the user both wants and does not want the technology. The user wants what the technology gives but does not want the limits, the transformations that a technologically extended body implies. There is a fundamental ambivalence toward the very human creation of our own earthly tools.

The ambivalence that can arise concerning technics is a reflection of one kind upon the *essential ambiguity* that belongs to technologies in use. But this ambiguity, I shall argue, has its own distinctive shape. Embodiment relations display an essential magnification/reduction structure which has been suggested in the instrumentation examples. Embodiment relations simultaneously magnify or amplify and reduce or place aside what is experienced through them.

The sight of the mountains of the moon, through all the transformational power of the telescope, removes the moon from its setting in the expanse of the heavens. But if our technologies were only to replicate our immediate and bodily experience, they would be of little use and ultimately of little interest. A few absurd examples might show this:

In a humorous story, a professor bursts into his club with the announcement that he has just invented a reading machine. The machine scans the pages, reads them, and perfectly reproduces them. (The story apparently was written before the invention of photocopying. Such machines might be said to be "perfect reading machines" in actuality.) The problem, as the innocent could see, was that this machine leaves us with precisely the problem we had prior to its invention. To have reproduced through mechanical "reading" all the books in the world leaves us merely in the library.

A variant upon the emperor's invisible clothing might work as well. Imagine the invention of perfectly transparent clothing through which we might technologically experience the world. We could see through it, breathe through it, smell and hear through it, touch through it. Indeed, it effects no changes of any kind, since it is *perfectly* invisible. Who would bother to pick up such clothing (even if the presumptive wearer could find it)? Only by losing some invisibility—say, with translucent coloring—would the garment begin to be usable and interesting. For here, at least, fashion would have been invented—but at the price of losing total transparency—by becoming that through which we relate to an environment.

Such stories belong to the extrapolated imagination of fiction, which stands in contrast to even the most minimal actual embodiment relations, which in their material dimensions simultaneously extend and reduce, reveal and conceal.

In actual human-technology relations of the embodiment sort, the transformational structures may also be exemplified by variations: In optical technologies, I have already pointed out how spatial significations change in observations through lenses. The entire gestalt changes. When the apparent size of the moon changes, along with it the apparent position of the observer changes. Relativistically, the moon is brought "close"; and equivalently, this optical near-distance applies to both the moon's appearance and my bodily sense of position. More subtly, every dimension of spatial signification also changes. For example, with higher and higher magnification, the well-known phenomenon of depth, instrumentally mediated as a "focal plane," also changes. Depth diminishes in optical near-distance.

A related phenomenon in the use of an optical instrument is that it transforms the spatial significations of vision in an instrumentally focal way. But my seeing without instrumentation is a full bodily seeing—I see not just with my eyes but with my whole body in a unified sensory experience of things. In part, this is why there is a noticeable irreality to the apparent position of the observer, which only diminishes with the habits acquired through practice with the instrument. But the optical instrument cannot so easily transform the entire sensory gestalt. The focal sense that is magnified through the instrument is monodimensioned.

Here may be the occasion (although I am not claiming a cause) for a certain interpretation of the senses. Historians of perception have noted that, in medieval times, not only was vision not the supreme sense but sound and smell may have had greatly enhanced roles so far as the interpretation of the senses went. Yet in the Renaissance and even more exaggeratedly in the Enlightenment, there occurred the reduction to sight as the favored sense, and within sight, a certain reduction of sight. This favoritism, however, also carried implications for the other senses.

One of these implications was that each of the senses was interpreted to be clear and distinct from the others, with only certain features recognizable through a given sense. Such an interpretation impeded early studies in echo location.

In 1799 Lazzaro Spallanzani was experimenting with bats. He noticed not only that they could locate food targets in the dark but also that they could do so blindfolded. Spallanzani wondered if bats could guide themselves by their ears rather than by their eyes. Further experimentation, in which the bats' ears were filled with wax, showed that indeed they could not guide themselves without their ears. Spallanzani surmised that either bats locate objects through hearing or they had some sense of which humans knew nothing. Given the doctrine of separate senses and the identification of shapes and objects through vision alone, George Montagu and Georges Cuvier virtually laughed Spallanzani out of the profession.

This is not to suggest that such an interpretation of sensory distinction was due simply to familiarity with optical technologies, but the common experience of enhanced vision through such technologies was at least the standard practice of the time. Auditory technologies were to come later. When auditory technologies did become common, it was possible to detect the same amplification/reduction structure of the human-technology experience.

The telephone in use falls into an auditory embodiment relation. If the technology is good, I hear *you* through the telephone and the apparatus "withdraws" into the enabling background:

(I-telephone)-you

But as a monosensory instrument, your phenomenal presence is that of a voice. The ordinary multidimensioned presence of a face-to-face encounter does not occur, and I must at best imagine those dimensions through your vocal gestures. Also, as with the telescope, the spatial significations are changed. There is here an auditory version of visual near-distance. It makes little difference whether you are geographically near or far, none at all whether you are north or south, and none with respect to anything but your bodily relation to the instrument. Your voice retains its partly irreal near-distance, reduced from the full dimensionality of direct perceptual situations. This telephonic distance is different both from immediate face-to-face encounters and from visual or geographical distance as normally taken. Its distance is a mediated distance with its own identifiable significations.

While my primary set of variations is to locate and demonstrate the invariance of a magnification/reduction structure to any embodiment relation, there are also secondary and important effects noted in the histories of tech-nology. In the very first use of the telephone, the users were fascinated and intrigued by its auditory transparency. Watson heard and recognized Bell's *voice*, even though the instrument had a high ratio of noise to message. In short, the fascination attaches to magnification, amplification, enhancement. But, contrarily, there can be a kind of forgetfulness that equally attaches to the reduction. What is *revealed* is what excites; what is concealed may be forgotten. Here lies one secret for technological trajectories with respect to development. There are *latent telics* that occur through inventions.

Such telics are clear enough in the history of optics. Magnification provided the fascination. Although there were stretches of time with little technical progress, this fascination emerged from time to time to have led to compound lenses by Galileo's day. If some magnification shows the new, opens to what was poorly or not at all previously detected, what can greater magnification do? In our own time, the explosion of such variants upon magnification is dramatic. Electron enhancement, computer image enhancement, CAT and NMR internal scanning, "big-eye" telescopes—the list of contemporary magnificational and visual instruments is very long.

I am here restricting myself to what may be called a *horizontal* trajectory, that is, optical technologies that bring various micro- or macro-phenomena to vision through embodiment relations. By restricting examples to such phenomena, one structural aspect of embodiment relations may be pointed to concerning the relation to microperception and its Adamic context. While *what* can be seen has changed dramatically—Galileo's New World has now been enhanced by astronomical phenomena never suspected and by micro-phenomena still being discovered—there remains a strong phenomenological constant in *how* things are seen. All lenses and optical technologies of the sort being described bring what is to be seen into a normal bodily space and distance. Both the macroscopic and the microscopic appear within the same near-distance. The "image size" of galaxy or amoeba is the *same*. Such is the existential condition for visibility, the counterpart to the technical condition, that the instrument makes things visually present.

The mediated presence, however, must fit, be made close to my actual bodily position and sight. Thus there is a reference within the instrumental context to my face-to-face capacities. These remain primitive and central within the new mediational context. Phenomenological theory claims that for every change in what is seen (the object correlate), there is a noticeable change in how (the experiential correlate) the thing is seen.

In embodiment relations, such changes retain both an equivalence and a difference from non-mediated situations. What remains constant is the bodily focus, the reflexive reference back to my bodily capacities. What is seen must be seen from or within my visual field, from the apparent distance in which discrimination can occur regarding depth, etc., just as in face-to-face relations. But the range of what can be brought into this proximity is transformed by means of the instrument.

Let us imagine for a moment what was never in fact a problem for the history of instrumentation: If the "image size" of both a galaxy and an amoeba is the "same" for the observer using the instrument, how can we tell that one is macrocosmic and the other microcosmic? The "distance" between us and these two magnitudes, Pascal noted, was the same in that humans were interpreted to be between the infinitely large and the infinitely small.

What occurs through the mediation is not a problem *because our construction of the observation presupposes ordinary praxical spatiality*. We handle the paramecium, placing it on the slide and then under the microscope. We aim the telescope at the indicated place in the sky and, before looking through it, note that the distance is at least that of the heavenly dome. But in our imagination experiment, what if our human were *totally immersed* in a technologically mediated world? What if, from birth, all vision occurred only through lens systems? Here the problem would become more difficult. But in our distance from Adam, it is precisely the presumed difference that makes it possible for us to see both nakedly *and* mediately—and thus to be able to locate the difference—that places us even more distantly from any Garden. It is because we retain this ordinary spatiality that we have a reflexive point of reference from which to make our judgments.

The noetic or bodily reflexivity implied in all vision also may be noticed in a magnified way in the learning period of embodiment. Galileo's telescope had a small field, which, combined with early hand-held positioning, made it very difficult to locate any particular phenomenon. What must have been noted, however, even if not commented upon, was the exaggerated sense of bodily motion experienced through trying to fix upon a heavenly body—and more, one quickly learns something about the earth's very motion in the attempt to use such primitive telescopes. Despite the apparent fixity of the stars, the hand-held telescope shows the earth-sky motion dramatically. This magnification effect is within the experience of one's own bodily viewing.

This bodily and actional point of reference retains a certain privilege. All experience refers to it in a taken-for-granted and recoverable way. The bodily condition of the possibility for seeing is now twice indicated by the very situation in which mediated experience occurs. Embodiment relations continue to locate that privilege of my being here. The partial symbiosis that occurs in well-designed embodied technologies retains that motility which can be called expressive. Embodiment relations constitute one existential form of the full range of the human-technology field.

Hermeneutic Technics

Heidegger's hammer in use displays an embodiment relation. Bodily action through it occurs within the environment. But broken, missing, or mal-functioning, it ceases to be the means of praxis and becomes an obtruding *object* defeating the work project. Unfortunately, that negative derivation of objectness by Heidegger carries with it a block against understanding a second existential human-technology relation, the type of relation I shall term *hermeneutic*.

The term hermeneutic has a long history. In its broadest and simplest sense it means "interpretation," but in a more specialized sense it refers to *textual* interpretation and thus entails *reading*. I shall retain both these senses and take hermeneutic to mean a special interpretive action within the technological context. That kind of activity calls for special modes of action and perception, modes analogous to the reading process.

Reading is, of course, a reading of_____ ; and in its ordinary con-text, what fills the intentional blank is a text, something *written*. But all writing entails technologies. Writing has a product. Historically, and more ancient than the revolution brought about by such crucial technologies as the clock or the compass, the invention and development of writing was surely even more revolutionary than clock or compass with respect to human experience. Writing transformed the very perception and understanding we have of language. Writing is a technologically embedded form of language.

There is a currently fashionable debate about the relationship between speech and writing, particularly within current Continental philosophy. The one side argues that speech is primary, both historically and ontologically, and the other—the French School—inverts this relation and argues for the primacy of writing. I need not enter this debate here in order to note the *technological difference* that obtains between oral speech and the materially connected process of writing, at least in its ancient forms.

Writing is inscription and calls for both a process of writing itself, employing a wide range of technologies (from stylus for cuneiform to word processors for the contemporary academic), and other material entities upon which the writing is recorded (from clay tablet to computer printout). Writing is technologically mediated language. From it, several features of hermeneutic technics may be highlighted. I shall take what may at first appear as a detour into a distinctive set of human-technology relations by way of a phenomenology of reading and writing.

Reading is a specialized perceptual activity and praxis. It implicates my body, but in certain distinctive ways. In an ordinary act of reading, particularly of the extended sort, what is read is placed before or somewhat under one's eyes. We read in the immediate context from some miniaturized bird's-eye perspective. What is read occupies an expanse within the focal center of vision, and I am ordinarily in a somewhat rested position. If the object-correlate, the "text" in the broadest sense, is a chart, as in the navigational examples, what is represented retains a representational isomorphism with the natural features of the landscape. The chart represents the land- (or sea)scape and insofar as the features are isomorphic, there is a kind of representational "transparency." The chart in a peculiar way "refers" beyond itself to what it represents.

Now, with respect to the embodiment relations previously traced, such an isomorphic representation is both similar and dissimilar to what would be seen on a larger scale from some observation position (at bird's-eye level). It is similar in that the shapes on the chart are reduced representations of distinctive features that can be directly or technologically mediated in face-to-face or embodied perceptions. The reader can compare these similarities. But chart reading is also different in that, during the act of reading, the perceptual focus is the chart itself, a substitute for the landscape.

I have deliberately used the chart-reading example for several purposes. First, the "textual" isomorphism of a representation allows this first example of hermeneutic technics to remain close to yet differentiated from the perceptual isomorphism that occurs in the optical examples. The difference is at least perceptual in that one sees through the optical technology, but now one sees the chart as the visual terminus, the "textual" artifact itself.

Something much more dramatic occurs, however, when the representational isomorphism disappears in a printed text. There is no isomorphism between the printed word and what it "represents," although there is some kind of *referential* "transparency" that belongs to this new technologically embodied form of language. It is apparent from the chart example that the chart itself becomes the *object of perception* while simultaneously referring beyond itself to what is not immediately seen. In the case of the printed text, however, the referential transparency is distinctively different from technologically embodied perceptions. *Textual transparency is hermeneutic transparency, not perceptual transparency.*

Historically, textual transparency was neither immediate nor attained at a stroke. The "technology" of phonetic writing, which now is increasingly a world-wide standard, became what it is through a series of variants and a process of experimentation. One early form of writing was pictographic. The writing was still somewhat like the chart example; the pictograph retained a certain representational isomorphism with what was represented. Later, more complex ideographic writing (such as Chinese) was, in effect, a more abstract form of pictography.

Calligraphers have shown that even early phonetic writing followed a gradual process of formalizing and abstracting from a pictographic base. Letters often depicted a certain animal, the first syllable of whose name provided the sound for the letter in a simultaneous sound and letter. Built into such early phonetic writing was thus something like the way the alphabet is still taught to children: "C is for Cow." Most educated persons are familiar with the mixed form of writing, hieroglyphics. Although the writing is pictographic, not all pictographs stood for the entity depicted; some represented sounds (phonemes) (see fig. 6.1).

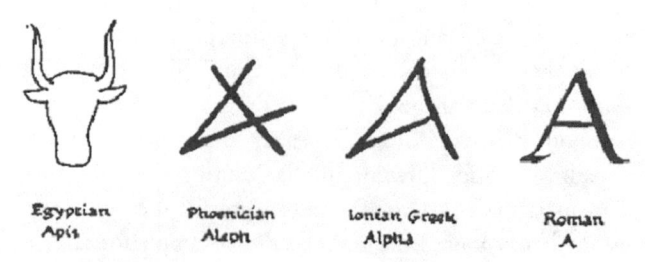

Figure 6.1. Caligraphy Progression. Drawing created by Don Ihde.

An interesting cross-cultural example of this movement from a very pictographic to a formalized and transformed ideographic writing occurs with Chinese writing. The same movement from relatively concrete representations in pictographs occurs through abbreviated abstractions—but in a different direction, non-phonetic and ideographic. Thus, for phonetic writing there is a double abstraction (from pictograph to letter and then reconstituting a small finite alphabet into represented spoken words), whereas the doubled abstraction of ideographic writing does not reconstitute to words as such, but to concepts.

In the most ancient Chinese writing in the period of the "Tortoise Shell Language" (prior to 2000 B.C.) and even in some cases through the later "Metal Language" period (2000–500 B.C.), if one is familiar with the objects as they occur within Chinese culture, one can easily detect the pictographic representation involved. For example, one can see that the ideograph for boat actually abstractly represents the sampan-type boats of the riverways (still in use). Similarly, in the ideograph for gate one can still recognize the uniquely Oriental-type gate in the drawing. The modern variants—related but more abstracted—have clearly lost that instant representational isomorphism (see fig. 6.2).

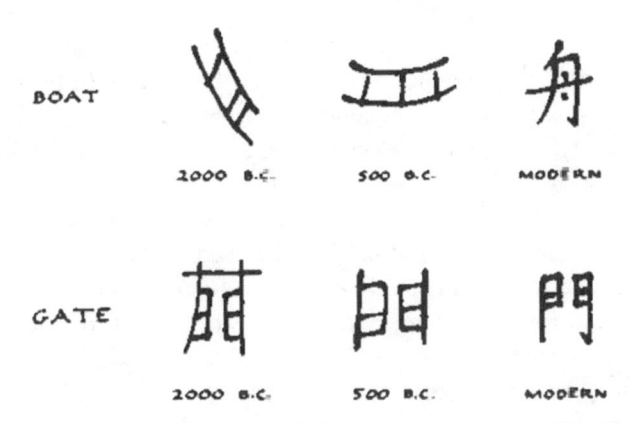

Figure 6.2. Boat and Gate Symbols. Drawing created by Don Ihde.

Implied in these transformations are changes of both technique and related technologies. Sergei Eisenstein, the film maker and one sensitive to such image technologies, has pointed to just such a transformation which arose out of the invention of the brush and India ink:

But then, by the end of the third century, the brush is invented. In the first century after the "joyous event" (A.D.)—paper. And, lastly, in the year 220—India ink.

A complete upheaval. A revolution in draughtmanship. And, after having undergone in the course of history no fewer than fourteen different styles of handwriting, the hieroglyph crystallized in its present form. The means of production (brush and India ink) determined the form.

The fourteen reforms had their way. As a result:

Figure 6.3. Horse Symbol. Drawing created by Don Ihde.

In the fierily cavorting hieroglyph *ma* (a horse) it is already impossible to recognize the features of the dear little horse sagging pathetically in its hindquarters, in the writing style of Ts'ang Chieh, so well-known from ancient Chinese bronzes. (Eisenstein 1949, 29)

If this is an accurate portrayal of the evolution of writing, it follows something like a Husserlian origin-of-geometry trajectory. The trajectory was from the more concrete to the greater degrees of abstraction, until virtually all "likeness" to origins disappeared. In this respect, writing only slowly approximated speech.

Once attained, like any other acquisition of the lifeworld, writing could be read and understood in terms of its unique linguistic transparency. Writing becomes an embodied hermeneutic technics. Now the descriptions may take a different shape. What is referred to is referred by the text and is referred to *through* the text. What now presents itself is the "world" of the text.

This is not to deny that all language has its unique kind of transparency. Reference beyond itself, the capacity to let something become present through language, belongs to speech as well. But here the phenomenon being centered upon is the new embodiment of language in writing. Even more thematically, the concern is for the ways in which writing as a "technology" transforms experiential structures.

Linguistic transparency is what makes present the *world* of the text. Thus, when I read Plato, Plato's "world" is made present. But this presence is a *hermeneutic* presence. Not only does it occur *through* reading, but it takes its shape in the interpretive context of my language abilities. His world is linguistically mediated, and while the words may elicit all sorts of imaginative and perceptual phenomena, it is through language that such phenomena occur. And while such phenomena may be strikingly rich, they do not appear as word-like.

We take this phenomenon of reading for granted. It is a sedimented acquisition of the literate lifeworld and thus goes unnoticed until critical reflection isolates its salient features. It is the same with the wide variety of hermeneutic technics we employ.

The movement from embodiment relations to hermeneutic ones can be very gradual, as in the history of writing, with little-noticed differentiations along the human-technology continuum. A series of wide-ranging variants upon readable technologies will establish the point. First, a fairly explicit example of a readable technology: Imagine sitting inside on a cold day. You look out the window and notice that the snow is blowing, but you are toasty warm in front of the fire. You can clearly "see" the cold in Merleau-Ponty's pregnant sense of perception—but you do not actually *feel* it. Of course, you could, were you to go outside. You would then have a full face-to-face verification of what you had seen.

But you might also see the thermometer nailed to the grape arbor post and *read* that it is 28°F. You would now "know" how cold it was, but you still would not feel it. To retain the full sense of an embodiment relation, there must also be retained some isomorphism with the felt sense of the cold—in this case, tactile—that one would get through face-to-face experience. One could invent such a technology; for example, some conductive material could be placed through the wall so that the negative "heat," which is cold, could be felt by hand. But this is not what the thermometer does.

Instead, you read the thermometer, and in the immediacy of your reading you *hermeneutically* know that it is cold. There is an instantaneity to such reading, as it is an already constituted intuition (in phenomenological terms). But you should not fail to note that perceptually what you have seen is the dial and the numbers, the thermometer "text." And that text has hermeneutically delivered its "world" reference, the cold.[1]

Such constituted immediacy is not always available. For instance, although I have often enough lived in countries where Centigrade replaces Fahrenheit, I still must translate from my intuitive familiar language to the less familiar one in a deliberate and self-conscious hermeneutic act. Immediacy, however, is not the test for whether the relation is hermeneutic. A hermeneutic relation mimics sensory perception insofar as it is also a kind of seeing as _____ ; but it is a referential seeing, which has as its immediate perceptual focus seeing the thermometer.

Now let us make the case more complex. In the example cited, the experiencer had both embodiment (seeing the cold) and hermeneutic access to the phenomenon (reading the thermometer). Suppose the house were hermetically sealed, with no windows, and the only access to the weather were through the thermometer (and any other instruments we might include). The hermeneutic character of the relation becomes more obvious. I now clearly have to know how to read the instrumentation and from this reading knowledge get hold of the "world" being referred to.

This example has taken actual shape in nuclear power plants. In the Three Mile Island incident, the nuclear power system was observed only through instrumentation. Part of the delay that caused a near meltdown was *misreadings* of the instruments. There was no face-to-face, independent access to the pile or to much of the machinery involved, nor could there be.

An intentionality analysis of this situation retains the mediational position of the technology:

I-technology-world
(engineer-instruments-pile)

The operator has instruments between him or her and the nuclear pile. But—and here, an essential difference emerges between embodiment and hermeneutic relations—what is immediately perceived is the instrument panel itself. It becomes the object of my microperception, although in the special sense of a hermeneutic transparency, I *read* the pile through it. This situation calls for a different formalization:

I-(technology-world)

The parenthesis now indicates that the immediate *perceptual* focus of my experience *is* the control panel. I read through it, but this reading is now dependent upon the semi-opaque connection between the instruments and the referent object (the pile). This *connection* may now become enigmatic.

In embodiment relations, what allows the partial symbiosis of myself and the technology is the capacity of the technology to become perceptually transparent. In the optical examples, the glass-maker's and lens-grinder's arts must have accomplished this end if the embodied use is to become possible. Enigmas which may occur regarding embodiment-use transparency thus may occur within the parenthesis of the embodiment relation (see fig. 6.4).

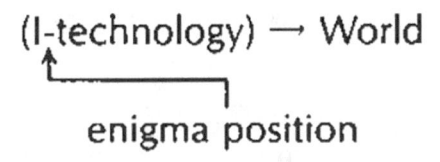

Figure 6.4. Enigma Position 1. Drawing created by Don Ihde.

(This is not to deny that once the transparency is established, thus making microperception clear, the observer may still fail, particularly at the macroperceptual level. For the moment, however, I shall postpone this type of interpretive problem.) It would be an oversimplification of the history of lens-making were not problems of this sort recognized. Galileo's instrument not only was hard to look through but was good only for certain "middle range" sightings in astronomical terms (it did deliver the planets and even some of their satellites). As telescopes became more powerful, levels, problems with chromatic effects, diffraction effects, etc., occurred. As Ian Hacking has noted, "Magnification is worthless if it magnifies two distinct dots into one big blur. One needs to resolve the dots into two distinct images. . . . It is a matter of diffraction. The most familiar example of diffraction is the fact that shadows of objects with sharp boundaries are fuzzy. This is a consequence of the wave character of light."[2] Many such examples may be found in the history of optics, technical problems that had to be solved before there could be any extended reach within embodiment relations. Indeed, many of the barriers in the development of experimental science can be located in just such limitations in instrumental capacity.

Here, however, the task is to locate a parallel difficulty in the emerging new human-technology relation, hermeneutic relations. The location of the technical problem in hermeneutic relations lies in the *connector* between the instrument and the referent. Perceptually, the user's visual (or other) terminus is *upon* the instrumentation itself. To read an instrument is an analogue to reading a text. But if the text does not correctly refer, its reference object or its world cannot be present. Here is a new location for an enigma (see fig. 6.5).

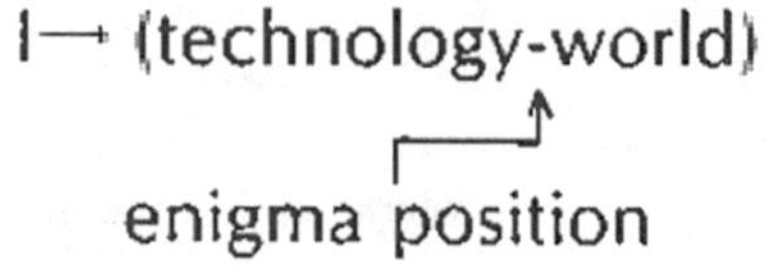

I—→ (technology-world)
⌐——↑
enigma position

Figure 6.5. Enigma Position 2. Drawing created by Don Ihde.

While breakdown may occur at any part of the relation, in order to bring out the graded distinction emerging between embodiment and hermeneutic relations, a short pathology of connectors might be noted.

If there is nothing that impedes my direct perceptual situation with respect to the instrumentation (in the Three Mile Island example, the lights remain on, etc.), interpretive problems in reading a strangely behaving "text" at least occur in the open; but the technical enigma may also occur within the text-referent relation. How could the operator tell if the instrument was malfunctioning or that to which the instrument refers? Some form of *opacity* can occur within the technology-referent pole of the relation. If there is some independent way of verifying which aspect is malfunctioning (a return to unmediated face-to-face relations), such a breakdown can be easily detected. Both such occurrences are reasons for instrumental redundancy. But in examples where such independent verification is not possible or untimely, the opacity would remain.

Let us take a simple mechanical connection as a borderline case. In shifting gears on my boat, there is a lever in the cockpit that, when pushed forward, engages the forward gear; upward, neutral; and backwards, reverse. Through it, I can ordinarily feel the gear change in the transmission (embodiment) and recognize the simple hermeneutic signification (forward for forward) as immediately intuitive. Once, however, on coming in to the dock at the end of the season, I disengaged the forward gear—and the propeller continued to drive the boat forward. I quickly reversed—and again the boat continued. The hermeneutic significance had failed; and while I also felt a difference in the way the gear lever felt, I did not discover until later that the clasp that retained the lever itself had corroded, thus preventing any actual shifting at all. But even at this level there can be opacity within the technology-object relation.

The purpose of this somewhat premature pathology of human-technology relations is not to cast a negative light upon hermeneutic relations in contrast to embodiment ones but rather to indicate that there are different locations where perceptual and human-technology relations interact. Normally, when the technologies work, the technology-world relation would retain its unique hermeneutic transparency. But if the I-(technology-world) relation is far enough along the continuum to identify the relation as a hermeneutic one, the intersection of perceptual-bodily relations with the technology changes.

Readable technologies call for the extension of my hermeneutic and "linguistic" capacities *through* the instruments, while the reading itself retains its bodily perceptual location as a relation *with* or *towards* the technology. What is emerging here is the first suggestion of an emergence of the technology as "object" but without its negative Heideggerian connotation. Indeed, the type of special capacity as a "text" is a condition for hermeneutic transparency.

The transformation made possible by the hermeneutic relation is a transformation that occurs precisely through *differences* between the text and what is referred to. What is needed is a particular set of textually clear perceptions that "reduce" to that which is immediately readable. To return to the Three Mile Island example, one problem uncovered was that the instrument panel design was itself faulty. It did not incorporate its dials and gauges in an easily readable way. For example, in airplane instrument panel design, much thought has been given to pattern recognition, which occurs as a perceptual gestalt. Thus, in a four-engined aircraft, the four dials indicating r.p.m. will be coordinated so that a single glance will indicate which, if any, engine is out of synchronization. Such technical design accounts for perceptual structures.

There is a second caution concerning the focus upon connectors and pathology. In all the examples I have used to this point, the hermeneutic technics have involved material connections. (The thermometer employs a physical property of a bimetallic spring or mercury in a column; the instrument panel at TMI employs mechanical, electrical, or other material connections; the shift lever, a simple mechanical connection.) If reading does not employ any such material connections, it might seem that its referentiality is essentially different, yet not even all technological connections are strictly material. Photography retains representational isomorphism with the object, yet does not "materially" connect with its object; it is a minimal beginning of action at a distance.

I have been using contemporary or post-scientific examples, but non-material hermeneutic relations do not obtain only for contemporary humans. As existential relations, they are as "old" as post-Garden humanity. Anthropology and the history of religions have long been familiar with a wide variety of shamanistic praxes which fall into the pattern of hermeneutic technics. In what may at first seem a somewhat outrageous set of examples, note the various "reading" techniques employed in shamanism. The reading of animal entrails, of thrown bones, of bodily marks—all are hermeneutic techniques. The patterns of the entrails, bones, or whatever are taken to *refer* to some state of affairs, instrumentally or textually.

Not only are we here close to a familiar association between magic and the origins of technology suggested by many writers, but we are, in fact, closer to a wider hermeneutic praxis in an intercultural setting. For that reason, the very strangeness of the practice must be critically examined. If the throwing of bones is taken as a "primitive" form of medical diagnosis—which does play a role in shamanism—we might conclude that it is indeed a poor form of hermeneutic relations. What we might miss, however, is that the entire gestalt of what is being diagnosed may differ radically from the other culture and ours.

It may well be that as a focused form of diagnosis upon some particular bodily ailment (appendicitis, for example), the diagnosis will fail. But since one important element in shamanism is a wider diagnosis, used particularly as the occasion of locating certain communal or social problems, it may work better. The sometimes socially contextless emphasis of Western medicine upon a presumably "mechanical" body may overlook precisely the context which the shaman so clearly recognizes. The entire gestalt is different and differently focused, but in both cases there are examples of hermeneutic relations.

In our case, the very success of Western medicine in certain diseases is due to the introduction of technologies into the hermeneutic relation (fever/thermometer; blood pressure/manometer, etc.) The point is that hermeneutic relations are as commonplace in traditional and ancient social groups as in ours, even if they are differently arranged and practiced.

By continuing the intentionality analysis I have been following, one can now see that hermeneutic relations vary the continuum of human-technology-world relations. Hermeneutic relations maintain the general mediation position of technologies within the context of human praxis towards a world, but they also change the variables within the human-technology-world relation. A comparative formalism may be suggestive:

General intentionality relations
Human-technology-world
Variant A: embodiment relations
(I-technology) → world
Variant B: hermeneutic relations
I → (technology world)

While each component of the relation changes within the correlation, the overall shapes of the variants are distinguishable. Nor are these matters of simply how technologies are experienced.

Another set of examples from the set of optical instruments may illustrate yet another way in which instrumental intentionalities can follow new trajectories. Strictly embodiment relations can be said to work best when there is both a transparency and an isomorphism between perceptual and bodily action within the relation. I have suggested that a trajectory for development in such cases may often be a horizontal one. Such a trajectory not only follows greater and greater degrees of magnification but also entails all the difficulties of a technical nature that go into allowing what is to be seen as though by direct vision. But not all optical technologies follow this strategy. The introduction of hermeneutic possibilities opens the trajectory into what I shall call *vertical* directions, possibilities that rely upon quite deliberate hermeneutic transformations.

It might be said that the telescope and microscope, by extending vision while transforming it, remained *analogue* technologies. The enhancement and magnification made possible by such technologies remain visual and transparent to ordinary vision. The moon remains recognizably the moon, and the microbe—even if its existence was not previously suspected—remains under the microscope a beastie recognized as belonging to the animate continuum. Here, just as the capacity to magnify becomes the foreground phenomenon to the background phenomenon of the reduction necessarily accompanying the magnification, so the similitude of what is seen with ordinary vision remains central to embodiment relations.

Not all optical technologies mediate such perceptions. In gradually moving towards the visual "alphabet" of a hermeneutic relation, deliberate variations may occur which enhance previously undiscernible *differences*:

1. Imagine using spectacles to correct vision, as previously noted. What is wanted is to *return* vision as closely as possible to ordinary perception, not to distort or modify it in any extreme micro- or macroperceptual direction. But now, for snowscapes or sun on the water or desert, we modify the lenses by coloring or polarizing them to cut glare. Such a variation transforms *what* is seen in some degree. Whether we say the polarized lens removes glare or "darkens" the landscape, what is seen is now clearly different from what may be seen through untinted glasses. This difference is a clue which may open a new *telic direction* for development.

2. Now say that somewhere, sometime, someone notes that certain kinds of tinting reveal unexpected results. Such is a much more complex technique now used in infrared satellite photos. (For the moment, I shall ignore the fact that part of this process is a combined embodiment and hermeneutic relation.) If the photo is of the peninsula of Baja California, it will remain recognizable in shape. Geography, whatever depth and height representations, etc., remain but vary in a direction different from any ordinary vision. The infrared photo enhances the difference between vegetation and non-vegetation beyond the limits of any isomorphic color photography. This difference corresponds, in the analogue example, to something like a pictograph. It simultaneously leaves certain analogical structures there and begins to modify the representation into a different, non-perceived "representation."

3. Very sophisticated versions of still representative but non-ordinary forms of visual recognition occur in the new heat-sensitive and light-enhanced technologies employed by the military and police. Night scopes which enhance a person's heat radiation still look like a person but with entirely different regions of what stands out and what recedes. In high-altitude observations, "heat shadows" on the ground can indicate an airplane that has recently had its engines running compared to others which have not. Here visual technologies bring into visibility what was not visible, but in a distinctly now perceivable way.

4. If now one takes a much larger step to spectrographic astronomy, one can see the acceleration of this development. The spectrographic picture of a star no longer "resembles" the star at all. There is no point of light, no disk size, no spatial isomorphism at all—merely a band of differently colored rainbow stripes. The naive reader would not know that this was a picture of a star at all—the reader would have to know the language, the alphabet, that has coded the star. The astronomer-hermeneut does know the language and "reads" the visual "ABCs" in such a way that he knows the chemical composition of the star, its internal makeup, rather than its shape or external configuration. We are here in the presence of a more fully hermeneutic relation, the star mediated not only instrumentally but in a transformation such that we must now thematically *read* the result. And only the informed reader can do the reading.

There remains, of course, the *reference* to the star. The spectrograph is *of* Rigel or *of* Polaris, but the individuality of the star is now made present hermeneutically. Here we have a beginning of a special transformation of perception, a transformation which deliberately enhances differences rather than similarities in order to get at what was previously unperceived.

5. Yet even the spectrograph is but a more radical transformation of perception. It, too, can be transformed by a yet more radical *hermeneutic* analogue to the *digital* transformation which lies embedded in the preferred quantitative praxis of science. The "alphabet" of science is, of course, mathematics, a mathematics that separates itself by yet another hermeneutic step from perception embodied.

There are many ways in which this transformation can and does occur, most of them interestingly involving a particular act of *translation* that often goes unnoticed. To keep the example as simple as possible, let us assume *mechanical* or *electronic* "translation." Suppose our spectrograph is read by a machine that yields not a rainbow spectrum but a set of numbers. Here we would arrive at the final hermeneutic accomplishment, the transformation of even the analogue to a digit. But in the process of hermeneuticization, the "transparency" to the object referred to becomes itself enigmatic. Here more explicit and thematic interpretation must occur.

Hermeneutic relations, particularly those utilizing technologies that permit vertical transformations, move away from perceptual isomorphism. It is the *difference* between what is shown and how something is shown which is informative. In a hermeneutic relation, the world is first transformed into a text, which in turn is read. There is potentially as much flexibility within hermeneutic relations as there are in the various uses of language. Emmanuel Mournier early recognized just this analogical relationship with language: "The machine as implement is not a simple material extension of our members. It is of another order, an annex to our language, an auxiliary language to mathematics, a means of penetrating, dissecting and revealing the secret of things, their implicit intentions, their unemployed capacities" (Mournier 1951, 195).

Through hermeneutic relations we can, as it were, *read* ourselves into any possible situation without being there. In science, in contrast to literature, what is important is that the reading retain *some* kind of reference or hermeneutic transparency to what is there. Perhaps that is one reason for the constant desire to reverse what is read back towards what may be perceived. In this reversal, contemporary technologically embodied science has frequently derived what might be called *translation technologies*. I mention two in passing:

(a) Digital processes have become *de rigueur* within the perceptual domain. The development of pictures from space probes is such a double translation process. The photograph of the surface of Venus is a technological analogue to human vision. It at least is a field display of the surface, incorporating the various possible figures and contrasts that would be seen instantaneously in a visual gestalt—but this holistic result cannot be transmitted in this way by the current technologies. Thus it is "translated" into a digital code, which can be transmitted. The "seeing" of the instrument is broken down into a series of digits that are radiographically transmitted to a receiver; then they are reassembled into a spatter pattern and enhanced to reproduce the photograph taken millions of miles away. It would be virtually impossible for anyone to read the digits and tell what was to be seen; only when the linear text of the digits has been retranslated back into the span of an instantaneous visual gestalt can it be seen that the rocks on Venus are or are not like those on the moon. Here the analogues of perception and language are both utilized to extend vision beyond the earth.

(b) The same process is used audially in digital recordings. Once again, the double translation process takes place and sound is reduced to digital form, reproduced through the record, and translated back into an auditory gestalt.

Digital and analogue processes blur together in certain configurations. Photos transmitted as points of black on a white ground and reassembled within certain size limits are perceptually gestalted; we see Humphrey Bogart, not simply a mosaic of dots. (Pointillism did the same in painting, although in color. So-called concrete poetry employs the same crossover by placing the words of the poem in a visual pattern so the poem may be both read and seen as a visual pattern.)

Such translation and retranslation processes are clearly transformations from perceptually gestalted phenomena into analogues of writing (serial translation and retranslation processes are clearly transformations from perceptual gestalt phenomena into analogues of writing serial transmissions along a "line," as it were), which are then re-translatable into perceptual gestalts.

I have suggested that the movement from embodiment relations to hermeneutic ones occurs along a human-technology continuum. Just as there are complicated, borderline cases along the continuum from fully haired to bald men, there are the same less-than-dramatic differences here. I have highlighted some of this difference by accenting the bodily-perceptual distinctions that occur between embodiment and hermeneutic relations. This has allowed the difference in perceptual and hermeneutic transparencies to stand out.

There remain two possible confusions that must be clarified before moving to the next step in this phenomenology of technics. First, there is a related sense in which perception and interpretation are intertwined. Perception is primitively already interpretational, in both micro- and macrodimensions. To perceive is already "like" reading. Yet reading is also a specialized act that receives both further definition and elaboration within literate contexts. I have been claiming that one of the distinctive differences between embodiment and hermeneutic relations involves perceptual position, but in the broader sense, interpretation pervades both embodiment and hermeneutic action.

A second and closely related possible confusion entails the double sense in which a technology may be used. It may be used simultaneously both as something *through* which one experiences and as something to which one relates. While this is so, the doubled relation takes shapes in embodiment different from those of hermeneutic relations. Return to the simple embodiment relation illustrated in wearing eyeglasses. *Focally*, my perceptual experience finds its directional aim *through* the lenses, terminating my gaze upon the object of vision; but as a fringe phenomenon, I am simultaneously aware of (or can become so) the way my glasses rest upon the bridge of my nose and the tops of my ears. In this fringe sense, I am aware *of* the glasses, but the focal phenomenon is the perceptual transparency that the glasses allow.

In cases of hermeneutic transparency, this doubled role is subtly changed. Now I may carefully read the dials within the core of my visual field and attend to them. But my reading is simultaneously a reading through them, although now the terminus of reference is not necessarily a perceptual object, nor is it, strictly speaking, perceptually present. While the type of transparency is distinct, it remains that the purpose of the reading is to gain hermeneutic transparency.

Both relations, however, at optimum, occur within the familiar acquisitional praxes of the lifeworld. Acute perceptual seeing must be learned and, once acquired, occurs as familiarly as the act of seeing itself. For the accomplished and critical reader, the hermeneutic transparency of some set of instruments is as clear and as immediate as a visual examination of some specimen. The peculiarity of hermeneutic transparency does not lie in either any deliberate or effortful accomplishment of interpretation (although in learning any new text or language, that effort does become apparent). That is why the praxis that grows up within the hermeneutic context retains the same sense of spontaneity that occurs in simple acts of bodily motility. Nevertheless, a more distinctive presence *of* the technology appears in the example. My awareness of the instrument panel is both stronger and centered more focally than the fringe awareness of my eyeglasses frames, and this more distinct awareness is essential to the optimal use of the instrumentation.

In both embodiment and hermeneutic relations, however, the technology remains short of full objectiveness or *otherwise*. It remains the means through which something else is made present. The negative characterization that may occur in breakdown pathologies may return. When the technology in embodiment position breaks down or when the instrumentation in hermeneutic position fails, what remains is an obtruding, and thus negatively derived, object.

Both embodiment and hermeneutic relations, while now distinguished, remain basic existential relations between the human user and the world. There is the danger that my now-constant and selective use of scientific instrumentation could distort the full impact of the existential dimension. Prior to moving further along the human-technology-world continuum, I shall briefly examine a very different set of instrumental examples. The instrumentation in this case will be *musical instrumentation.*

In the most general sense, it should be easy to see that the use of musical instrumentation, in performance, falls into the same configurations as do scientific instruments:

I—musical instrument—world
I—scientific instrument—world

But the praxical context is significantly changed. If scientific or knowledge-developing praxis is constrained by the need to have a referential terminus within the world, the musical praxis is not so constrained. Indeed, if there is a terminus, it is a reference not so much to some thing or region of the environment as to the production of a musical event within that environment. The "musical object" is whatever sound phenomenon occurs through the performance upon the instrument. Musical sounds are produced, *created.* Whereas in the development of scientific instrumentation the avoidance of phenomena that would be artifacts of the instrument rather than of its referent are to be avoided or reduced as much as possible, the very discovery and enhancement of such instrumental artifacts may be a positive phenomenon in making music. There are interesting and significant differences in these two praxical contexts, but for the moment, I shall restrict myself to a set of observations about the similarities in the intentionality structures of both scientific and musical instrumentation.

It should be obvious that a very large use of musical instrumentation falls clearly into the embodiment relation pattern. The player picks up the instrument (having learned to embody it) and expressively produces the desired music:

Player-instrument-sound

In embodiment cases, the sound-making instrument will be partially symbiotically embodied:

(Player-instrument)-sound

Second, the previously noted amplification/reduction structure also occurs here. If our player is a trombonist, the "buzz" his lip vibrations produce can be heard without any instrument but, once amplified and transformed *through* the trombone, occur as the musical sound distinctive to the human-instrument pairing. Equally immediately, at least within the complex of contemporary instrumentation, one may detect that nothing like a restriction to human sound as such belongs to the contemporary musical context. Isomorphism to human sound, while historically playing a significant cultural role, now occupies only one dimension of musical sound.

This history, however, is interesting. There have been tendencies in Western musical history to restrict to or at least to develop precisely along horizontally variant ways. The restriction of musical sound to actual human voices (certain Mennonite sects do not allow any musical instrumentation, and all hymn singing is done a cappella) is a form of this tendency. Instrumentation that mimics or actually amplifies vocal sounds and their ranges is another example: woodwinds, horns, organs (even to the organ stop titles which are usually voice analogues)—all are ancient instruments that often deliberately followed a kind of vocal isomorphism. Medieval music was often doubly constrained. Not only must the music remain within the range of human similitude, but even the normatively controlled harmonics and chant lines were religio-culturally constrained. Later, one could detect a much more vocal model to much Italian (Renaissance through Baroque) music in contrast to a more instrumentally oriented model in German music.

The implicit valuational model of the human voice was also reflected in the music history of the West by the ranking of instruments by *expressivity*, with those instruments thought most expressive—the violin, for example— rated more highly than those farther from the vocal model.

The difference between embodiment and hermeneutic relations appears within this context as well. While embodiment relations in the most general existential sense need not be strictly constrained by isomorphism, hermeneutic variants occur very quickly along the musical spectrum. The piano retains little vocal isomorphism; yet when played, it falls into the embodiment relation, is expressive of the individual style and attainment of the performer, etc. Farther along the continuum, computer-produced music clearly occurs much more fully within the range of hermeneutic relations, in some cases with the emergence of random-sound generation very close to the sense of otherness, which will characterize the next set of relations where the technology emerges as *other*.

Instrumental music, as technics, may go in either embodiment or hermeneutic directions. It may develop its instrumentation in both vertical and horizontal trajectories. In either direction there are recognizable clear, technological transformations. If the Western "bionic" model of much early music was voice, in Andean music it was bird song (both in melody and in sound quality produced by breathy wood flutes). Contrarily, percussion instrumentation (drum music and communication) was, from the outset, a movement in a vertical and thus more hermeneutic direction. This exploration of possibility trees in horizontal and vertical directions belongs to the realm of musical praxis as much as to scientific, but is without any referentiality to a natural world.

The result of technological development in musical technics is also suggestively different from its result in scientific praxis. The "world" produced musically through all the technical adumbrations is *not* that suggested either by the new philosophy of science or by a Heideggerian philosophy of technology. The closest analogy to the notion of standing reserve (resource well) that the musical "world" might take is that the realm of all possible sound may be taken and/or transformed musically. But the acoustical resources of musical technics are utilized through the creative sense of *play* which pervades musical praxis. The "musical object" is a created object, but its creation is not constrained by the same imperatives of scientific praxis. Yet the materialization of musical sound *through* instrumentation remains a fully human technological form of action.

What can be glimpsed in this detour into musical instrumentation is that while the human-technology structures are parallel with those found within scientific instrumentation, the "world" created does not at all imply the same reduction to what has been claimed as the unique Western view of the domination of nature. Here, then, is an opening to a different possible trajectory of development.

Alterity Relations

Beyond hermeneutic relations there lie *alterity relations*. The first suggestions of such relations, which I shall characterize as relations to or with a technology, have already been suggested in different ways from within the embodiment and hermeneutic contexts. Within embodiment relations, were the technology to intrude upon rather than facilitate one's perceptual and bodily extension into the world, the technology's objectness would necessarily have appeared negatively. Within hermeneutic relations, however, there emerged a certain positivity to the objectness of instrumental technologies. The bodily-perceptual focus *upon* the instrumental text is a condition of its own peculiar hermeneutic transparency. But what of a positive or presentential sense of relations with technologies? In what phenomenological senses can a technology be *other*?

The analysis here may seem strange to anyone limited to the habits of objectivist accounts, for in such accounts technologies as objects usually come first rather than last. The problem for a phenomenological account is that objectivist ones are non-relativistic and thus miss or submerge what is distinctive about human-technology relations.

A naive objectivist account would likely begin with some attempt to circumscribe or define technologies by object characteristics. Then, what I have called the technical properties of technologies would become focal. Some combination of physical and material properties would be taken to be definitional. (This is an inherent tendency of the standard nomological positions such as those of Bunge and Hacking.) The definition will often serve a secondary purpose by being stipulative: only those technologies that are obviously dependent upon or strongly related to contemporary scientific and industrial productive practices will count.

This is not to deny that objectivist accounts have their own distinctive strengths. For example, many such accounts recognize that technological or "artificial" products are different from the simply found object or the natural object. But the submergence of the human-technology relation remains hidden, since either object may enter into praxis and both will have their material, and thus limited, range of technical usability within the relation. Nor is this to deny that the objectivist accounts of types of technologies, types of organization, or types of designed purposes should be considered. But the focus in this first program remains the phenomenological derivation of the set of human-technology relations.

There is a tactic behind my placing alterity relations last in the order of focal human-technology relations. The tactic is designed, on the one side, to circumvent the tendency succumbed to by Heidegger and his more orthodox followers to see the otherness of technology only in negative terms or through negative derivations. The hammer example, which remains paradigmatic for this approach, is one that derives objectness from breakdown. The broken or missing or malfunctioning technology could be discarded. From being an obtrusion it could become *junk*. Its objectness would be clear—but only partly so. Junk is not a focal object of use relations (except in certain limited situations). It is more ordinarily a background phenomenon, that which has been put out of use.

Nor, on the other side, do I wish to fall into a naively objectivist account that would simply concentrate upon the material properties of the technology as an object of knowledge. Such an account would submerge the relativity of the intentionality analysis, which I wish to preserve here. What is needed is an analysis of the positive or presentential senses in which humans relate to technologies as relations to or with technologies, to technology-as-other. It is this sense which is included in the term "alterity."

Philosophically, the term "alterity" is borrowed from Emmanuel Levinas. Although Levinas stands within the traditions of phenomenology and hermeneutics, his distinctive work, *Totality and Infinity*, was "anti-Heideggerian." In that work, the term "alterity" came to mean the radical difference posed to any human by another human, an *other* (and by the ultimately other, God). Extrapolating radically from within the tradition's emphasis upon the non-reducibility of the human to either objectness (in epistemology) or as a means (in ethics), Levinas poses the otherness of humans as a kind of *infinite* difference that is concretely expressed in an ethical, face-to-face encounter.

I shall retain but modify this radical Levinasian sense of human otherness in returning to an analysis of human-technology relations. How and to what extent do technologies become other or, at least, *quasi-other?* At the heart of this question lie a whole series of well-recognized but problematic interpretations of technologies. On the one side lies the familiar problem of anthropomorphism, the personalization of artifacts. This range of anthropomorphism can reach from serious artifact-human analogues to trivial and harmless affections for artifacts.

An instance of the former lies embedded in much AI research. To characterize computer "intelligence" as human-like is to fall into a peculiarly contemporary species of anthropomorphism, however sophisticated. An instance of the latter is to find oneself "fond" of some particular technofact as, for instance, a long-cared-for automobile which one wishes to keep going and which may be characterized by quite deliberate anthropomorphic terms. Similarly, in ancient or non-Western cultures, the role of sacredness attributed to artifacts exemplifies another form of this phenomenon.

The religious object (idol) does not simply "represent" some absent power but is endowed with the sacred. Its aura of sacredness is spatially and temporally present within the range of its efficacy. The tribal devotee will defend, sacrifice to, and care for the sacred artifact. Each of these illustrations contains the seeds of an alterity relation.

A less direct approach to what is distinctive in human-technology alterity relations may perhaps better open the way to a phenomenologically relativistic analysis. My first example comes from a comparison to a technology and to an animal "used" in some practical (although possibly sporting) context: the spirited horse and the spirited sports car.

To ride a spirited horse is to encounter a lively animal other. In its pre- or nonhuman context, the horse has a life of its own within the environment that allowed this form of life. Once domesticated, the horse can be "used" as an "instrument" of human praxis—but only to a degree and in a way different from counterpart technologies; in this case, the "spirited" sports car.

There are, of course, analogues which may at first stand out. Both horse and car give the rider/driver a magnified sense of power. The speed and the experience of speed attained in riding/driving are dramatic extensions of my own capacities. Some prominent features of embodiment relations can be found analogously in riding/driving. I experience the trail/road through horse/car and guide/steer the mediating entity under way. But there are equally prominent differences. No matter how well trained, no horse displays the same "obedience" as the car. Take malfunction: in the car, a malfunction "resists" my command—I push the accelerator, and because of a clogged gas line, there is not the response I expected. But the animate resistance of a spirited horse is more than such a mechanical lack of response—the response is more than malfunction, it is *dis*obedience. (Most experienced riders, in fact, prefer spirited horses over the more passive ones, which might more nearly approximate a mechanical obedience.) This life of the other in a horse may be carried much further—it may live without me in the proper environment; it does not need the *deistic* intervention of turning the starter to be "animated." The car will not shy at the rabbit springing up in the path any more than most horses will obey the "command" of the driver to hit the stone wall when he is too drunk to notice. The horse, while approximating some features of a mediated embodiment situation, never fully enters such a relation in the way a technology does. Nor does the car ever attain the sense of animation to be found in horseback riding. Yet the analogy is so deeply embedded in our contemporary consciousness (and perhaps the lack of sufficient experience with horses helps) that we might be tempted to emphasize the similarities rather than the differences.

Anthropomorphism regarding the technology on the one side and the contrast with horseback riding on the other point to a first approximation to the unique type of otherness that relations to technologies hold. Technological otherness is a *quasi-otherness*, stronger than mere objectness but weaker than the otherness found within the animal kingdom or the human one; but the phenomenological derivation must center upon the positive experiential aspects outlining this relation.

In yet another familiar phenomenon, we experience technologies as *toys* from childhood. A widely cross-cultural example is the spinning top. Prior to being put into use, the top may appear as a top-heavy object with a certain symmetry of design (even early tops approximate the more purely functional designs of streamlining, etc.), but once "deistically" animated through either stick motion or a string spring, the now spinning top appears to take on a life of its own. On its tip (or "foot") the top appears to defy its top-heaviness and gravity itself. It traces unpredictable patterns along its pathway. It is an object of *fascination*.

Note that once the top has been set to spinning, what was imparted through an embodiment relation now exceeds it. What makes it fascinating is this property of quasi-animation, the life of its own. Also, of course, once "automatic" in its motion, the top's movements may be entered into a whole series of possible contexts. I might enter a game of warring tops in which mine (suitably marked) represents me. If I-as-top am successful in knocking down the other tops, then this game of hermeneutics has the top winning for me. Similarly, if I take its quasi-autonomous motion to be a hermeneutic predictor, I may enter a divination context in which the path traced or the eventual point of stoppage indicates some fortune. Or, entering the region of scientific instrumentation, I may transform the top into a gyroscope, using its constancy of direction within its now-controlled confines as a better-than-magnetic compass. But in each of these cases, the top may become the focal center of attention as a quasi-other to which I may relate. Nor need the object of fascination carry either an embodiment or hermeneutic referential transparency.

To the ancient and contemporary top, compare briefly the fascination that occurs around video games. In the actual use of video games, of course, the embodiment and hermeneutic relational dimensions are present. The joystick that embodies hand and eye coordination skills extends the player into the displayed field. The field itself displays some hermeneutic context (usually either some "invader" mini-world or some sports analogue), but this context does not refer beyond itself into a worldly reference.

In addition to these dimensions, however, there is the sense of *interacting with* something other than me, the technological *competitor*. In competition there is a kind of dialogue or exchange. It is the quasi-animation, the quasi-otherness of the technology that fascinates and challenges. I must beat the machine or it will beat me.

In each of the cases mentioned, features of technological alterity have shown themselves. The quasi-otherness, the quasi-autonomy which appears in the toy or the game is a variant upon the technologies that have fascinated Western thinkers for centuries, the *automaton*.

The most sophisticated Greek (and similarly, Chinese) technologies did not appear in practical or scientific contexts so often as in game or theatrical ones. (War contexts, of course, have always employed advanced technologies.) Within these contexts, automatons were devised. From rediscovered treatises by Hero of Alexandria on pneumatics and hydraulics (which had in the second century B.C. already been used for humorous applications), the Renaissance builders began to construct various automata. The applications of Hero had been things like automatically opening temple doors and artificial birds that sang through steam whistles. In the Renaissance reconstructions, automata became more complex, particularly in fountain systems:

> The water garden of the Villa d'Este, built in 1550 at Tivoli, outside Rome, for the son of Lucrezia Borgia [was the best known]. The slope of the hill was used to supply fountains and dozens of grottos where water-powered figures moved and played and spouted. . . . The Chateau Merveilleux of Helbrun . . . is full of performing figures of men and women where fountains turn on and off unexpectedly or, operating in the intricate and quite amazing theatre of puppets, run by water power. (Burke 1978, 106)

The rage for automata was later to develop in a number of directions from music machines, of which the Deutsches Museum in Munich has a grand collection, to Vaucanson's automated duck which quacked, ate, drank, and excreted (Burke 1978, 107). Much later, automation techniques were used in more practical contexts, although versions of partially automated looms for textiles did begin to appear in the eighteenth century (Vaucanson, the maker of the automated duck, invented the holed cylinder that preceded the punch-card system of the Jacquard loom).

Nor should the clock be exempted from this glance at automata fascination. The movements of the heavens, of the march of life and death, and of the animated figures on the clocks of Europe were other objects of fascination that seemed to move "autonomously." The superficial aspects of automation, the semblance of the animate and the similitude of the human and animal, remained the focus for even more serious concerns with automatons. That which is more "like" us seemed to center the fascination and make the alterity more quasi-animate.

Fascination may hide what is reductive in technological selectivities. But it may also hide, doubly, a second dimension of an instrumental intentionality, its possible dissimilarity direction, which may often prove in the longer run the more interesting trajectory of development. Yet semblance usually appears to be the first focus.

It was this *semblance* which became a worry for Modern (seventeenth and eighteenth century) Philosophy. Descartes's famous doubts also utilize the popular penchant for automata. In seeking to prove that it is the mind alone and not the eyes that know things, he argues:

> I should forthwith be disposed to conclude that the wax is known by the act of sight and not by the intuition of the mind alone were it not for the analogous instance of human beings passing on in the street below, as observed from a window. In this case, I do not fail to say that I see the men themselves, just as I say that I see the wax; and yet, what do I see from the window beyond hats and cloaks that might cover artificial machines, whose motions might be determined by springs? (Descartes 1953, 92)

This can-I-be-fooled-by-a-cleverly-conceived-robot argument was to have an exceedingly long history, even into the precincts of contemporary analytic philosophies.

Were Descartes to become a contemporary of current developments in the attempt to mimic animal and human motions by automata, he might well rethink his illustration. Not only spring-run automata but also the most sophisticated computer-run automata look mechanical. These most sophisticated computer-run automata have difficulty maneuvering in anything like a lifelike motion. As Dreyfus has pointed out and as would be confirmed by many current researchers, bodily motion is perhaps harder to imitate than certain "mental" activities such as calculating.

To follow only the inclination towards similitude, however, is to reduce what may be learned from our relations with technologies. The current state of the art in AI research, for example, while having been partially freed from its earlier fundamentalistic state, remains primarily within the aim of creating similarities with human intelligence or modeling what are believed to be analogues to our intelligence. Yet it might well be that the *differences* that emerge from computer experimentation may be more informative or, at least, as informative as the similitudes.

There are what I shall call technological *intentionalities* that emerge from many technologies. Let us engage in a pseudo-Cartesian, imaginative construction of a humanoid robot, within the limits of easily combinable and available technologies, to take account of the similarity/difference structures which may be displayed. I shall begin with the technology's "perceptions" of sensory equipment: What if the robot were to hear? The inventor, perhaps limited by a humanist's budget, could install an omnidirectional microphone for ears. We could check upon what our robot would "hear" by adding a cassette player for a recorded "memory" of its "hearing." What is heard would turn out to be very differently structured, to have a very different form of intentionality than what any human listener would hear.

Assume that our robot is attending a university lecture in a large hall and is seated, as a shy student might be, near the rear. Given the limits of the mentioned technology, what would be heard would fail to have either the foreground/background pattern of human listening or the selective elimination of noise that even ordinary listening displays. The robot's auditory memory, played back, would reveal something much more like a sense-data auditory world than the one we are familiar with. The lecturer's voice, though recorded and within low limits perhaps detectable, would often be buried under the noise and background sounds that are selectively masked by human listening. For other purposes, precisely this differently structured technological intentionality could well be useful and informative. Such a different auditory selectivity could perhaps give clues to better architectural dampening of sounds precisely because what is repressed in human listening here stands out. In short, there is "truth" to be found in both the similarity and the difference that technological intentionalities reveal.

A similar effect could be noted with respect to the robot's vision. Were its eyes to be made of television equipment and the record or memory of what it has seen displayed on a screen, we would once again note the flatness of its visual field. Depth phenomena would be greatly reduced or would disappear. Although we have become accustomed to this flat field in watching television, it is easy to become reaware of the lack of depth between the baseball pitcher and the batter upon the screen. The technological shape of intentionality differs significantly from its human counterpart.

The fascination with human or animate similitude within the realm of alterity relations is but another instance of the types of fascination pervading our relations with technologies. The astonishment of Galileo at what he saw through the telescope was, in effect, the location of similitude within embodiment use. The magnification was the magnification *of* human visual capacity and remained within the range of what was familiarly visible. The horizontal trajectory of magnification that can more and more enhance vision is a trajectory along an already familiar praxis.

With the examples of fascination with automata, the fascination also remains within the realm of the familiar, now in a kind of mirror phenomenon for humans and the technology. Of all the animals in the earth's realm, it seems that the human ones are those who can prolong this fascination the most intensely. Paul Levinson, in an examination of the history of media technologies, has argued that there are three stages through which technologies pass. The first is that of technology as toy or novelty. The history of film technology is instructive: "The first film makers were not artists but tinkers. . . . "Their goal in making a movie was not to create beauty but to display a scientific curiosity." A survey of the early "talkies" like *The Jazz Singer*, first efforts in animation such as Disney's "Laugh-O-Gram" cartoons, and indeed the supposed debut of the motion picture in *Fred Ott's Sneeze* supports [this thesis] itself" (Levinson 1985, 163). The same observation could be made about much invention. But once taken more seriously, novelty can be transformed into a second stage, according to Levinson: that of technology as mirror of reality. This too happened in the history of film. Following the early curiosities at the onset of the film industry, the introduction of the Lumieres' presentation of "actualities" were, in part, fascinating precisely through the magnification/reduction selectivities that film technologies produce through unique film intentionalities. Examples could be as mundane as "workers leaving a factory, a baby's meal, and the famous train entering the station." What made such cinemas vérités dramatic were "in this case, a real train chugging into a real station, at an angle such that the audience could almost believe the train was chugging at *them*" (Levinson 1985, 165).

This mirror of life, like the automaton, is not isomorphic with non-technological experience but is technologically transformed with the various effects that exaggerate or enhance some effects while simultaneously reducing others. Levinson is quite explicit in his analysis concerning the ways newly introduced technologies also enhance this development:

> The growth of film from gimmick to replicator was apparently in large part dependent upon a new technological component. . . . The "toy" film played to individuals who peeked into individual kinetoscopes; but the "reality" film reached out to mass audiences, who viewed the reality-surrogate in group theatres. The connection between mass audiences and reality simulation, moreover, was no accident. Unlike the perception of novelties, which is inherently subjective and individualized, reality perception is a fundamentally objective, group process. (Levinson 1985, 167)

Although the progression of the analysis here moves from embodiment and hermeneutic relations to alterity ones, the interjection of film or cinema examples is of suggestive interest. Such technologies are transitional between hermeneutic and alterity phenomena. When I first introduced the notion of hermeneutic relations, I employed what could be called a "static" technology: writing. The long and now ancient technologies of writing result in fixed texts (books, manuscripts, etc., all of which, barring decay or destruction, remain stable in themselves). With film, the "text" remains fixed only in the sense that one can repeat, as with a written text, the seeing and hearing of the cinema text. But the mode of presentation is dramatically different. The "characters" are now animate and theatrical, unlike the fixed alphabetical characters of the written text. The dynamic "world" of the cinema-text, while retaining many of the functional features of writing, also now captures the semblance of real-time, action, etc. It remains to be "read" (viewed and heard), but the object-correlate necessarily appears more "life-like" than its analogue—written text. This factor, naively experienced by the current generations of television addicts, is doubtless one aspect in the problems that emerge between television watching habits and the state of reading skills. James Burke has pointed out that "the majority of the people in the advanced industrialized nations spend more time watching television than doing anything else beside work" (Burke 1978, 5). The same balance of time use also has shown up in surveys regarding students. The hours spent watching television among college and university students, nationally, are equal to or exceed those spent in doing homework or out-of-class preparation.

Film, cinema, or television can, in its hermeneutic dimension, refer in its unique way to a "world." The strong negative response to the Vietnam War was clearly due in part to the virtually unavoidable "presence" of the war in virtually everyone's living room. But films, like readable technologies, are also *presentations*, the focal terminus of a perceptual situation. In that emergent sense, they are more dramatic forms of perceptual immediacy in which the presented display has its own characteristics conveying quasi-alterity. Yet the engagement with the film normally remains short of an engagement with an *other*. Even in the anger that comes through in outrage about civilian atrocities or the pathos experienced in seeing starvation epidemics in Africa, the emotions are not directed to the screen but, indirectly, through it, in more appropriate forms of political or charitable action. To this extent there is retained a hermeneutic reference elsewhere than at the technological instrument. Its quasi-alterity, which is also present, is not fully focal in the case of such media technologies.

A high-technology example of breakdown, however, provides yet another hint at the emergence of alterity phenomena. Word processors have become familiar technologies, often strongly liked by their users (including many philosophers who fondly defend their choices, profess knowledge about the relative abilities of their machines and programs, etc.). Yet in breakdown, this quasi-love relationship reveals its quasi-hate underside as well. Whatever form of "crash" may occur, particularly if some fairly large section of text is involved, it occasions frustration and even rage. Then, too, the programs have their idiosyncrasies, which allow or do not allow certain movements; and another form of human-technology competition may emerge. (Mastery in the highest sense most likely comes from learning to program and thus overwhelm the machine's previous brainpower. "Hacking" becomes the game-like competition in which an entire system is the alterity correlate.) Alterity relations may be noted to emerge in a wide range of computer technologies that, while failing quite strongly to mimic bodily incarnations, nevertheless display a quasi-otherness within the limits of linguistics and, more particularly, of logical behaviors. Ultimately, of course, whatever contest emerges, its sources lie opaquely with other humans as well but also with the transformed technofact, which itself now plays a more obvious role within the overall relational net.

I have suggested that the computer is one of the stronger examples of a technology which may be positioned within alterity relations. But its otherness remains a quasi-otherness, and its genuine usefulness still belongs to the borders of its hermeneutic capacities. Yet in spite of this, the tendency to fantasize its quasi-otherness into an authentic otherness is pervasive. Romanticizations such as the portrayal of the emotive, speaking "Hal" of the movie *2001: A Space Odyssey*, early fears that the "brain power" of computers would soon replace human thinking, fears that political or military decisions will not only be informed by but also made by computers—all are symptoms revolving around the positing of otherness to the technology.

These romanticizations are the alterity counterparts to the previously noted dreams that wish for total embodiment. Were the technofact to be genuinely an other, it would both be and not be a *technology*. But even as quasi-other, the technology falls short of such totalization. It retains its unique role in the human-technology continuum of relations as the medium of transformation, but as a recognizable medium.

The wish-fulfillment desire occasioned by embodiment relations—the desire for a fully transparent technology that would *be* me while at the same time giving me the powers that the use of the technology makes available—here has its counterpart fantasy, and this new fantasy has the same internal contradiction: It both reduces or, here, extrapolates the technology into that which is not a technology (in the first case, the magical transformation is *into me*; in this case, *into the other*), and at the same time, it desires what is not identical with me or the other. The fantasy is for the transformational effects. Both fantasies, in effect, deny technologies playing the roles they do in the human-technology continuum of relations; yet it is only on the condition that there be some detectable differentiation within the relativity that the unique ways in which technologies transform human experience can emerge.

In spite of the temptation to accept the fantasy, what the quasi-otherness of alterity relations does show is that humans may relate positively or presententially to technologies. In that respect and to that degree, technologies emerge as focal entities that may receive the multiple attentions humans give the different forms of the other. For this reason, a third formalization may be employed to distinguish this set of relations:

$$I \rightarrow \text{technology-(-world)}$$

I have placed the parentheses thusly to indicate that in alterity relations there may be, but need not be, a relation through the technology to the world (although it might well be expected that the *usefulness* of any technology will necessarily entail just such a referentiality). The world, in this case, may remain context and background, and the technology may emerge as the foreground and focal quasi-other with which I momentarily engage.

This disengagement of the technology from its ordinary-use context is also what allows the technology to fall into the various disengaged engagements which constitute such activities as play, art, or sport.

A first phenomenological itinerary through direct and focal human-technology relations may now be considered complete. I have argued that the three sets of distinguishable relations occupy a continuum. At the one extreme lie those relations that approximate technologies to a quasi-me (embodiment relations). Those technologies that I can so take into my experience that through their semi-transparency they allow the world to be made immediate thus enter into the existential relation which constitutes my self. At the other extreme of the continuum lie alterity relations in which the technology becomes quasi-other, or technology "as" other to which I relate. Between lies the relation with technologies that both mediate and yet also fulfill my perceptual and bodily relation with technologies, hermeneutic relations. The variants may be formalized thus:

Human-technology-World Relations
Variant 1, Embodiment Relations
(Human-technology) $\rightarrow$ World
Variant 2, Hermeneutic Relations
Human $\rightarrow$ (technology-World)
Variant 3, Alterity Relations
Human $\rightarrow$ technology-(-World)

Although I have characterized the three types of human-technology relations as belonging to a continuum, there is also a sense in which the elements within each type of relation are differently distributed. There is a *ratio* between the objectness of the technology and its transparency in use. At the extreme height of embodiment, a background presence of the technology may still be detected. Similarly but with a different ratio, once the technology has emerged as a quasi-other, its alterity remains within the domain of human invention through which the world is reached. Within all the types of relations, technology remains artifactual, but it is also its very artifactual formation which allows the transformations affecting the earth and ourselves.

All the relations examined heretofore have also been focal ones. That is, each of the forms of action that occur through these relations have been marked by an implicated self-awareness. The engagements through, with, and to technologies stand within the very core of praxis. Such an emphasis, while necessary, does not exhaust the role of technologies nor the experiences of them. If focal activities are central and foreground, there are also fringe and background phenomena that are no more neutral than those of the foreground. It is for that reason that one final foray in this phenomenology of technics must be undertaken. That foray must be an examination of technologies in the background and at the horizons of human-technology relations.

Background Relations

With background relations, this phenomenological survey turns from attending to technologies in a foreground to those which remain in the background or become a kind of near-technological environment itself. Of course, there are discarded or no-longer-used technologies, which in an extreme sense occupy a background position in human experience—junk. Of these, some may be recuperated into non-use but focal contexts such as in technology museums or in the transformation into junk art. But the analysis here points to specifically functioning technologies which ordinarily occupy background or field positions.

First, let us attend to certain individual technologies designed to function in the background—automatic and semiautomatic machines, which are so pervasive today—as good candidates for this analysis. In the mundane context of the home, lighting, heating, and cooling systems, and the plethora of semiautomatic appliances are good examples. In each case, there is some necessity for an instant of deistic intrusion to program or set the machinery into motion or to its task. I set the thermostat; then, if the machinery is high-tech, the heating/ cooling system will operate independently of ongoing action. It may employ time-temperature changes, external sensors to adjust to changing weather, and other cybernetic operations. (While this may function well in the home situation, I remain amused at the still-primitive state of the art in the academic complex I occupy. It takes about two days for the system to adjust to the sudden fall and spring weather changes, thus making offices which actually have opening windows—a rarity—highly desirable.) Once operating, the technology functions as a barely detectable background presence; for example, in the form of background noise, as when the heating kicks in. But in operation, the technology does not call for focal attention.

Note two things about this human-technology relation: First, the machine activity in the role of background presence is not displaying either what I have termed a transparency or an opacity. The "withdrawal" of this technological function is phenomenologically distinct as a kind of "absence." The technology is, as it were, "to the side." Yet as a present absence, it nevertheless becomes part of the experienced field of the inhabitant, a piece of the immediate environment.

Somewhat higher on the scale of semiautomatic technologies are task-oriented appliances that call for explicit and repeated deistic interventions. The washing machine, dryer, microwave, toaster, etc., all call for repeated programming and then for dealing with the processed product (wash, food, etc.). Yet like the more automated systems, the semiautomatic machine remains in the background while functioning.

In both systems and appliances, however, one also may detect clues to the ways in which background relations texture the immediate environment. In the electric home, there is virtually a constant hum of one sort or the other, which is part of the technological texture. Ordinarily, this "white noise" may go unnoticed, although I am always reassured that it remains part of fringe awareness, as when guests visit my mountain home in Vermont. The inevitable comment is about the silence of the woods. At once, the absence of background hum becomes noticeable.

Technological texturing is, of course, much deeper than the layer of background noise which signals its absent presence. Before turning to further implications, one temptation which could occur through the too-narrow selection of contemporary examples must be avoided. It might be thought that only, or predominantly, the high-technology contemporary world uses and experiences technologies as backgrounds. That is not the case, even with respect to automated or semiautomatic technologies.

The scarecrow is an ancient "automated" device. Its mimicry of a human, with clothes flapping in the breeze, is a specifically designed automatic crow scarer, made to operate in the absence of humans. Similarly, in ancient Japan there were automated deer scarers, made of bamboo tubes, pivoted on a pin and placed so that a waterfall or running stream would slowly fill the tube. When it is full enough, the device would trip and its other end strike a sounding board or drum, the noise of which would frighten away any marauding deer. We have already noted the role automation plays in religious rituals (prayer wheels and worship representations thought to function continuously).

Interpreted technologically, there are even some humorous examples of "automation" to be found in ancient religious praxes. The Hindu prayer windmill "automatically" sends its prayers when the wind blows; and in the ancient Sumerian temples there were idols with large eyes at the altars (the gods), and in front of them were smaller, large-eyed human statues representing worshipers. Here was an ancient version of an "automated" worship. (Its contemporary counterpart would be the joke in which the professor leaves his or her lecture on a tape recorder for the class—which students could also "automatically" hear, by leaving their own cassettes to tape the master recording.)

While we do not often conceptualize such ancient devices in this way, part of the purpose of an existential analysis is precisely to take account of the identity of function and of the "ancientness" of all such existential relations. This is in no way to deny the differences of context or the degree of complexity pertaining to the contemporary, as compared to the ancient, versions of automation.

Another form of background relation is associated with various modalities of the technologies that serve to insulate humans from an external environment. Clothing is a borderline case. Clothing clearly insulates our bodies from temperature, wind, and other external weather phenomena that could become dangerous to life; but clothing experienced is borderline with embodiment relations, for we do feel the external environment through clothing, albeit in a particularly damped-down mode. Clothing is not designed, in most cases, to be "transparent" in the way the previous instrument examples were but rather to have a certain opacity without restricting movement. Yet clothing is part of a fringe awareness in most of our daily activities (I am obviously not addressing fashion aspects of clothing here).

A better example of a background relation is a shelter technology. Although shelters may be found (caves) and thus enter untransformed into human praxis, most are constructed, as are most technological artifacts; but once constructed and however designed to insulate or account for external weather, they become a more field-like background phenomenon. Here again, human cultures display an amazing continuum from minimalist to maximalist strategies with respect to this version of a near-background.

Many traditional cultures, particularly in Southern Hemisphere areas, practice an essentially open shelter technology, perhaps with primarily a roof to keep off rain and sun. Such peoples frequently find distasteful such items as windows and, particularly, glassed windows. They do not wish to be too isolated or insulated from the elements. At the other extreme is the maximalist strategy, which most extremely wishes to totalize shelter technology into a virtual life-support system, autonomous and enclosed. I shall call this a technological cocoon.

A contemporary example of a near-cocoon is the nuclear submarine. Its crew lives inside, and the vessel is designed to remain at sea for prolonged periods, even underwater for long stretches of time. There are sophisticated recycling systems for waste, water, and air. Contact with the outside, obviously important in this case, is primarily through monitoring equally sophisticated hermeneutic devices (sonar, low-frequency radio, etc.). All ordinary duties take place in the cocoon-like interior. A multibillion-dollar projection to a greater degree of cocoonhood is the long-term space station now under debate.

Part of the very purpose of the space station is to experiment with creating a mini-environment, or artificial "earth," which would be totally technologically mediated. Yet contemporary high-tech suburban homes show similar features. Fully automated for temperature and humidity, tight air structures, some with glass that adjusts to glare, all such homes lie on the same trajectory of self-containment. But while these illustrations are uniquely high-technology textured, there remain, as before, differently contexted but similar examples from the past.

Totally enclosed spaces have frequently been associated with ritual and religious praxis. The Kiva of past southwestern native American cultures was dug deep into the ground, windowless and virtually sealed. It was the site for important initiatory and secret societies, which gathered into such ancient cocoons for their own purposes. The enclosure bespeaks different kinds of totalization.

What is common to the entire range of examples pointed to here is the position occupied by such technology, background position, the position of an absent presence as a part of or a total field of immediate technology.

In each of the examples, the background role is a field one, not usually occupying focal attention but nevertheless conditioning the context in which the inhabitant lives. There are, of course, great differences to be detailed in terms of the types of contexts which such background technologies play. Breakdown, again, can play a significant indexical role in pointing out such differences.

The involvement implications of contemporary, high-technology society are very complex and often so interlocked as to fall into major disruption when background technology fails. In 1985 Long Island was swept by Hurricane Gloria with massive destruction of power lines. Most areas went without electricity for at least a week, and in some cases, two. Lighting had to be replaced by older technologies (lanterns, candles, kerosene lamps), supplies for which became short immediately. My own suspicion is that a look at birth statistics at the proper time after this radical change in evening habits will reveal the same glitch which actually did occur during the blackouts of earlier years in New York.

Similarly, with the failure of refrigeration, eating habits had to change temporarily. The example could be expanded quite indefinitely; a mass purchase of large generators by university buyers kept a Minnesota company in full production for several months after, to be prepared the "next time." In contrast, while the same effects on a shorter-term basis were experienced in the grid-wide blackouts of 1965, I was in Vermont at my summer home, which is lighted by kerosene lamps and even refrigerated with a kerosene refrigerator. I was simply unaware of the massive disruption until the Sunday *Times* arrived. Here is a difference between an older, loose-knit and a contemporary, tight-knit system.

Despite their position as field or background relations, technologies here display many of the same transformational characteristics found in the previous explicit focal relations. Different technologies texture environments differently. They exhibit unique forms of non-neutrality through the different ways in which they are interlinked with the human lifeworld. Background technologies, no less than focal ones, transform the gestalts of human experience and, precisely because they are absent presences, may exert more subtle indirect effects upon the way a world is experienced. There are also involvements both with wider circles of connection and amplification/reduction selectivities that may be discovered in the roles of background relations; and finally, the variety of minimalist to maximalist strategies remains as open to this dimension of human-technology relations as each of the others.

Horizonal Phenomena

The limits of this first analysis are now close at hand. With the indirect and field aspects of background relations, there are also hints of horizonal phenomena, which mark the boundaries of a phenomenology. (It should be noted that the term "horizon" in phenomenology is a limit concept. Unlike the common English expression "to expand one's horizons," no such possibility exists here. The horizon is the limit beyond which the inquiry ceases to display its internal characteristics, and like a voyage at sea, the horizon always remains that distant boundary between sky and sea or land and sea. It never comes nearer.)

Horizons, however, may be indicated in a number of ways with respect to extreme fringe phenomena. They are of important and indicative use here, such that mention must be made prior to moving to a second program. In the entire preceding analysis, the experience of technology has always been such that there is some experientially recoverable difference between what is experienced and the experiencer. All transparencies have been noted to be quasi-transparencies, all alterities quasi-otherness, and even the absent presences of background phenomena are at least indirectly recoverable.

Such a positioning of technologies as belonging to a continuum of relations, which points to their essential artifactual properties within human praxis, has been a constant found in all variations. Yet the question of the extremities beyond which there is no recovery, where perhaps technologies cease to be technologies, remains intriguing. I shall try to locate just such phenomena within the spectrum of the explicit phenomenologies I have followed.

In the case of embodiment relations, are there embodiments which fall "below" recovery? Embodiment technologies are taken into self-experience but remain distinctive, at least in fringe or echo phenomena, as other than my self. But what of the secret desire to have technology become myself? Such desires are approximated in the bionic trajectories of contemporary technologies. A borderline example is a crowned tooth. Although there is little doubt that something is being "added" to my mouth during the dental procedure and there is a period of accommodation in which I experience the strangeness of the cap, after a time the tooth becomes almost totally embodied. Yet even here, there remains a less-distinct recoverable awareness of the difference in texture, the feeling of hot and cold, and the slipperiness that calls for special chewing gums if it is not to stick. But the artifactuality of the cap is, indeed, an extreme to recover as a relation. It comes close to "being me."

Less so are the various implants which, users report, are experienced as a different mode of being one's own body. Hip joints of stainless steel and teflon do allow the previously restricted walker to resume close-to-normal walking, but the discernible difference remains a fringe awareness of the bionically transformed individual. Yet such prosthetic implantations do allow a being-towards-the-world as a modified bodily being.

Such mechanical transplants do not yet reach the extremity of horizonal phenomena. An even more extreme set of examples arises from chemical transformations, or what I shall call edible technologies. The history of the birth-control pill is instructive in this case. Early users of the pill reported two results: They did experience bodily changes, in that period pains were experienced differently; and sometimes there were other side effects such as minor nausea. But as with the previously noted fascination with the amplifying transformations of all new technologies, most such side effects were repressed in favor of the exultation over a worry-free ability to engage in close-to-"natural" or pregnancy-free sexual relations.

Later, delayed side effects were associated with the pill (in some cases, elevated blood pressure, etc.; later, worries over cancers), but these could at best be indirectly experienced. The pill, once taken, functioned as a kind of internal background relation of the most extreme fringe type. As with all edible technologies, the "I am what I eat" phenomenon placed most effects at a distance or were delayed.

This very case, however, is instructive in a different way. Are edible technologies technologies at all? They are clearly technological products (of the chemistry industry) and are clearly technologically transformed entities; but insofar as they are absorbable into our very bodies, they, much more closely than our previous artifactual examples, are deeply embodied.

Even more ambiguously, the question of biological technology must be raised at this horizonal extreme. I refer here not to the popular issues revolving around whether animal life, through gene manipulation, can or should be patented but to the now literal reflexivity which biological technologies may have upon the transformation of human bodily life itself. A genetic transformation does display the same structural transformational features that the other human-technology relations display, but insofar as a genetic result becomes what we are, it dips below the level of at least the contrastive existential analysis taken here. This is not to say, however, that a hermeneutically enhanced analysis is precluded.

With biological technics, there is reached a new boundary between technology and life where the horizons of nature and artificiality are blurred. Yet while the ability to manipulate genetic material more precisely through identification of particular genes or strands is contemporary (a feat made possible by the instrumental ability to magnify the micro-levels of being, hence, conditioned by an essential possibility of technology) such biological technics are also ancient. Plant and animal breeding, hybridization, and other manipulations of wild products are as old as civilization and occurred independently in many parts of the world.

Furthermore, the blurring of the borderline life/artificial product is anticipated in the changes of attitude and use towards domestic plants and animals. They become clearly "use beings" under human control and for human use.

Horizons belong to the boundaries of the experienced environmental field. Like the "edges" of the visual field, they situate what is explicitly present, while as a phenomenon itself, horizons recede. And whether we refer to a kind of inner horizon (the fringes of embodiment) or the extremities of the external horizon (the ultimate form of texturing that a specific technological culture may take), the result is one of "atmosphere." This does not mean that the elements, particularly in terms of social phenomena, cannot be identified. For example, in contemporary situations, one atmospheric phenomenon is the vague fear or anxiety related to the possibility of nuclear threat. This fear is felt by children as well as parents and across at least all industrialized cultures. It is part of the texture of the technological culture that has an ultimate and universalizable destructive magnificational power. But this atmospheric social fear phenomenon is an effect of something else: the dream of technological totalization which marks one feature of maximalist technological cultures.

There are a number of ways in which nuclear fear might be historically analyzed. Obviously, destruction of one's enemies has been a human aim as ancient as humanity. Equally obviously, it has never been possible until the twentieth century to actualize this aim universally. Even Samson tires after thousands; but the potential, however quickly (full exchange of superpower weaponry) or slowly (nuclear winter), to eliminate the species is only possible for the present and the future. The inherent problem of a universal totalized solution is equally obvious—it is self-reflexive and applies to the killer as well as the killed. The political necessity to preserve such a universally genocidal means of "defense" takes the shape of attempts to deny self-reflexivity. Governments must find ways to convince populaces that there will be survivors (nuclear winter will be only autumn, etc.).

Yet the atmospheric nuclear fear is but the underside of a previous century's utopianism that a technological culture, once totalized, would be able to solve all basic social problems such as hunger, disease, a decent standard of living, and peaceful interchange between nations. If such hopes are reduced to single system utopianism in the late twentieth century, they nevertheless spring from the same mythical sources as the current fears.

Eve and the Spaceship

Perhaps what is needed at the end of this first itinerary is another fictive variation: to imagine the most extreme maximization of technology, for the other end of the continuum that began in the imagination of a non-technological Garden. This projection must aim at a totalized technological culture. The problems of even imagining such a state of existence are parallel to those found with the dream of technological innocence at the beginning, for we must project from our present situation in which there is only a presumption of totalization.

Yet there are clues to a trajectory of totalization from precisely this present. The imagination, science-fiction-like, would have to find technologically totalized components for each dimension of the existential situation of human beings in the world previously surveyed.

For World, we would have to have an artifactual "world," or self-contained environment, that was totally autonomous within some complex of technological conditions. Although we are no more able to point to such an artificial "world" than we are able to find any actual Garden, the concept of the technological cocoon is the instance of such a trajectory. It is even part of our fictional tradition and now a part of our economic debate (how many billions are we willing to spend on such an experiment?). At the present, such a situation would be temporarily lived in and clearly would be highly dependent upon earth stations, but the dream is for totalization. Yet approximations to fully enclosed mini-cocoons are, in fact, the recreational vehicles, the fully controlled building environments, and even certain kinds of condominiums we actually do inhabit. Insulated as much as possible from weather (unless it is pleasant) the projected artificial world is but an extension of present conditions. It is also a reflection of present desires.

At first, this somewhat ironic projection could be taken to hide some form of nature romanticism, a form which frequents even the widest space fantasies of many cinema and television series. Whereas the focal plot and visual effects are all of projected high-technology spaceships, alien beings, etc., the theme all too often is a quest for a lost earth or a new earth. Romanticism also belongs to utopian thought. But that is not the point; rather, I wish to point up that insofar as we are successful in actually constructing approximations to totalized or at least maximized technological cultures, the result is necessarily to replace the dominant regions of threat and problem from "nature" to precisely technological "culture."

In the last few years, the triplet—"Challenger," Bhopal, and Chernobyl— has become almost symbolic of this replacement of threat phenomenon. Each breakdown was felt and taken up as a peculiar late-twentieth-century event. Bhopal and Chernobyl, by far the largest catastrophes in terms of human impact, became equivalents of the past major natural disasters such as hurricanes or tidal waves, sweeping the world with their aftereffects. Fears of breakdown on a large scale become fears of technologically textured societies.

These events, which occupy worldwide news, are indicative of a subtle but clear replacement or at least equivalence of past threats largely from natural disasters. The same mix of wider and instant communication, even to the interinvolvement of media and event, occurs in the strange archaic/contemporary movements of much international terrorism. Revivals of religious fundamentalism, whose feuds now are spread with contemporary weaponry and publicized by contemporary media (from the tape-cassettes that spread Islamic fundamentalism underneath the controlled news media of the Shah, to the linkage of terrorist publicity, to the need for instant television coverage). This, too, is part of the substitution process.

Not all interlinkage, however, is symbolized by breakdown. Within the same tragedies were the internationalization of health care, with American bone marrow transplant experts helping Soviet patients and with rock stars stimulating a vast worldwide food transport program for Ethiopian starvation victims. The interlinking of media and event here is just as close as in the above negative examples.

If the technological cocoon is the microcosm for an artificial world in the dream of a totalized technological culture, its other components have also received imaginative treatment. If the cocoon is a replacement for a wider world, then the inhabitants must undergo similar extrapolative treatment.

When we turn to examples of the new technological Eves who inhabit our spaceship cocoons in science fiction—cinema or television form—they are frequently hybrids or imitations (androids or robots cleverly contrived, as Descartes imagined, to be mistaken for humans). The bionic being, whether woman or man, is a being conceived of with new technological parts that unlike all extant such parts, have improved upon the weakness of the replaced part. Such imaginings of hybrids, more powerful and more perfect than the previous merely biological beings, are reflections of precisely the technological dreaming earlier noted. The bionic beings have become perfect unions of technology and life such that they simply are and experience themselves as superbeings. The technology has ceased to be a technology and become a (more perfect) human.

Fictional androids and robots, however, are nearly perfect replicas or semblances of humans, although most plots eventually reveal them to be Cartesian-clever machines, usually with some hidden flaw. These entities, which in this scheme would be even more alterities than actual technofacts, begin to reveal the underside of fear that is directed at the aim of the dream itself. What if we were able to actually equal or exceed ourselves through our own inventions? Would our inventions then find us of no use or, worse, of use to them?

Take the imagination of technological perfecting in a different direction. What if we were technologically able to duplicate a human being, including all the foibles and warts? Would we not be in the perfect reading-machine situation? What sense would there be to a technology perfectly duplicative of something non-technological? At best, such an invention would satisfy the copy urges of the designer, as a perfect forgery would the forger.

But if the android were simultaneously both human and transformed, then it might be of interest. And here the fictional varieties abound: the perfect human slave, or the perfect human lover, or the perfect human strong person, etc. Yet here again we reach the contradiction of the dream in which the desire is both to have and not to have a technology, with its artifactual and selective essence. No human Eve would ever wish to be a "Stepford Wife."

Although the above fictive variations have been directed at the others who might cohabit within the cocoon, what has been revealed there applies in a similar way to a possible technological totalization of the self. Which Eve in the spaceship would voluntarily yield to a total technological transformation? The answer is, in contemporary life, somewhat ambiguous. The use of steroids by athletes is apparently widespread in spite of known risks (later muscle deterioration, high blood pressure, increased risk of heart failure). The use of cosmetic surgery, including implants, is fashionable. On a much more ancient level, the human uses of everything from makeup to initiatory masks and scarification are instrumentally applied transformations of the human body. Yet none of these practices comes close to the totalized variation, for beyond the above probably trivial pursuits, few Adams or Eves would willingly submit to heart, liver, or prosthetic transplants except when the loss of life or motility threatens.

That we should stop short of totalization is not itself strange; what is strange is the persistence of the modern Cartesian dream of the perfectability of the cleverly designed automaton as a substitute for the human being.

Dreams of Totalization

Within this first itinerary through human-technology relations there appears to be a vast anomaly. On the one side, an examination of human-technology relations reveals a deep continuity in the ways humans have experienced their technologies. These existential relations apply equally to ancient and modern, to simple and complex technologies. Insofar as a structural phenomenology aims at invariants, this is as it should be; yet this emphasis upon sameness leaves something out—particularly regarding what appears to contemporaries as the vast difference between all traditional and nonindustrial societies and our own.

In part, this impression is part of a tactic of this first program in human-technology analysis. I have tried to show that there is a continuity that spans both human history and cultures beyond the extravagant claims of those who would dissociate us from our pasts and our peers. Yet to disregard the disjuncture between today's maximalist technological cultures and those of minimalist or less-than-minimalist lifeforms would also be distorting.

Not that hints are lacking regarding these differences. One such theme, followed somewhat narrowly but purposely here, has been the emphasis upon the disjunctions made possible through modern instrumentation. Modern science, in my view, is distinctively different from all ancient science by virtue of its embodiment in instrumentation, whether of embodiment or hermeneutic types, for these have made possible views of the universe that both imaginatively and perceptually exceed and differ from all ancient cosmologies. Alfred North Whitehead has made the same point: "The reason we are on a higher imaginative level is not because we have a finer imagination but because we have better instruments. In science, the most important thing that has happened in the last forty years is the advance in instrumental design . . . a fresh instrument serves the same purpose as foreign travel; it shows things in unusual combinations. The gain is more than a mere addition; it is a transformation" (Whitehead 1963, 107).

That transformation is a quality disjunction that begins to occur with the rise of modern science but is greatly accelerated only in more recent times.

A second clue to a disjunction between ancient and contemporary times lies in the ways dreams of totalization are actualized in today's maximalist and yesterday's minimalist cultures. Indeed, the fact that humans have become a virtual geological force through the magnified technologies of recent time is the seat of one of our distinctive contemporary worries.

If industrial products are a major cause of the newly discovered ozone hole in the Antarctic as suspected, if the threatened elimination of a number of small animal species because of rain forest destruction (a percentage roughly equivalent in number to the past's large animal species extinction), if acid rain creating pollution threatens most higher-altitude northern hemisphere forests, then we have reason for this worry.

Yet while the intensity of the human-through-technology impact in today's world is clearly of a magnitude greater than in any historical time, it is not without precedent. J. Donald Hughes, in a remarkable if pessimistic book, *Ecology in Ancient Civilizations*, argued that regardless of cultural differences, each of the civilizations surrounding the Mediterranean produced the same results of (a) deforestation, (b) over-grazing by sheep and goats, and (c) irreversible erosion resulting in (d) today's arid climate all around the Mediterranean basin.

This result was similar whether we speak of Greeks, Hebrews, Romans, Phoenicians, or any of the other progenitors of our later European civilization. Ecologically speaking, Hughes provides an anti-romantic counterpart to Heidegger's glorification of the Greek temple: "Those who look at the Parthenon, that incomparable symbol of the achievements of an ancient civilization, often do not see its wider setting. Behind the Acropolis, the bare, dry mountains of Attica show their rocky bones against the blue Mediterranean sky, and the ruin of the finest temple built by the ancient Greeks is surrounded by the far vaster ruins of an environment which they desolated at the same time" (Hughes 1975, 1).

Even more ancient are the implications of humans in the extinctions of the large mammals of the late Ice Age. Or, to turn to a more recent and non-Western example, similar extinctions of all the large wingless birds on most Pacific islands occurred with the westward expansion of the Polynesian navigators of whom I have earlier spoken. Such results are clearly not the unique effects of either our cultural ancestors or our histories but lie deeper in the possibility structures of the human.

Do such trajectories arise from the dreams of totalization suggested here? Again, the answer must respect cultural ambiguity. To complete the first itinerary, I wish to contrast what I shall call two different dreams of totalization. I shall take as my examples our own technologically maximalist one and contrast it with one of the most technologically minimalist cultures known to me, that of the inland Australian Aborigines.

For illustrative purposes only—because I find deep flaws in such an interpretation—I shall assume that those critics of our historical development are correct in characterizing the dominant trend of our Technological (with the reified "T") culture as one whose dream of totalization is geared to bringing nature itself under the control of culture. If the purpose of technology is to control nature or to make it available to human purposes, then interpretations such as those of Lynn White, Jr., who saw in Western history a search for power arising out of religious consciousness (to subdue and dominate the earth), or that of Heidegger, who saw the revealing power of modern technology to have revealed the world as a "resource well" (my translation of Bestand), are correct.

If one then takes the inland Aboriginal cultures as a contrast, it might seem that here lies an exception to precisely the direction taken by our histories. So taken, the example is impressive: First, one must note that the ancestors of the Aboriginals arrived at least 40,000 years ago, as evidenced by artifacts found in the south of Australia. Second, they were largely isolated from external invasions and migrations (an exception apparently occurred in the north some three thousand years ago when the dingo was introduced; this non-neutral introduction of a wild dog led to the extinction of the marsupial wolf, one of the few marsupial predators). Thus, the Aboriginals may be said to be one of the world's longest-lasting isolated but continuous cultures.

That they survived at all in such arid conditions is impressive, but to have survived in such a way that their primary culture could be character-ized by what we might call "leisure activities" is even more noteworthy. In that context, it should be noted that very few hours of the day are spent gathering or preparing food. Much more time, relatively, is spent in their version of philosophical argument—times of storytelling and arguing over a religio-mythical system noted for its complexity. Celebratory activities take up much of the communal cultural time. Weddings, funerals, seasonal festivals were noted to last weeks on end at the time of the first European explorers. Artistic activity, too, was highlighted in both visual and auditory form. Arnheim Land's "X-ray" paintings and didjereedoo music are today valued by denizens of our postmodern culture.

Yet this long-lasting culture was clearly one of the most minimalist of technological cultures. Not that it lacked material artifacts altogether: There was a weapons collection, including the invention of both returning and non-returning boomerangs, spears (pre-Stone Age; tips were hardened wood), spear-throwing devices to amplify arm power, slings, etc., but not the bow and arrow found virtually worldwide in other ancient cultures.

And if the men's repertory of hunting weapons is to be matched by an equal set of technologies on the female-gathering side (and this more important food source), then the advanced basketry, adapted emu egg car-riers, digging, and other tools are also to be noted.

Within religious practice there were thuringas and masks and a wide variety of aesthetic-religious artifacts, including the famous bark paintings and cave drawings of "X-ray" animal depictions. Yet the spectrum is minimal by any comparison to any other non-migratory peoples' set of tools or manufactured objects.

"Technical" knowledge, however, was vast and complex—particularly with respect to food sources. As I began to study this culture, I was tempted to speculate that one central cultural axiom may well have been: "Everything is potentially food; the only question is how to make it so." The Aboriginals discovered or invented detoxification processes that sometimes took three steps, in contrast to many other cultures' single heat-by-boiling-or-baking or soaking techniques. The range of food sources included some thirty varieties of breads made from grass and weed seeds, delicacies of baked bats caught by children, and the hard-to-find witchetty grub (a five-inch-long grub inhabiting the root system of the witchetty tree). The grub is found by stomping the heel of one's foot around the base of the tree and listening to the echo-located hollow sounds.

Tales are told of the amazement of the Aboriginals at the first Europeans to try to cross the central deserts of Australia. Many of those folk died of thirst, desiccation, and exposure. The Aboriginal response was wonder at how, in the land of such plenty, anyone could be so stupid. For water, for example, all one had to do was to find the crack in the dried mud floor of a proper location and dig down some two feet to a membrane-enclosed, water-surrounded frog awaiting the next rains (which might be several years apart), drink the water, and eat the frog for a chaser!

Here, then, is a minimalist culture that of necessity learned to cope with a harsh environment. For the culture to succeed, however, there were prices to be paid. A second axiom, after maximizing the edible, must be one directed at minimizing population growth. Here, as expected, are means to be found: An unusual one is a form of "male birth control" in the practice of ritual subincision of adolescent male children in puberty rites. Subincision consists of a permanent hole made at the bottom base of the penis through the use of a thorn inserted during the initiation ceremony. Thereafter, a true "man" will have to hold this hole when urinating to mark him as a man. An unexpected side effect is that there will be a limited amount of sperm escaping during intercourse, thus effectively lowering the sperm count and hence lowering the probability of insemination. Of course, more drastic measures also had to be taken to keep the population down. Associated with the belief that no couple should have more than two children (who could be carried in flight; a third would theoretically have to be left behind), there is evidence of another familiar South Pacific means of birth control: infanticide.

Such prices are unlikely to be paid by those of us inhabiting our own cultures. Similarly, life expectancies, tolerance of weather extremes, and the like are not about to be traded by us for survival in such a lifeform. And the ultimate intercultural answer also seems visible among Aboriginals today as well—it is a culture largely subsumed within the fringes of a now two-century-old European dominant culture.

I shall not disguise a certain admiration for the human learning and adaptability that inland Aboriginals exemplified, particularly in relation to what I would call a religioethical system that functioned as a conservationist ethic. Two simple beliefs indicate how such an ethic functioned. One rule relates to the prohibition against killing any animal outside its sacred territory. The context for this rule is explained in terms not likely to be believed by us. It includes elements of totemism, in which the animals noted belong to a kin group (although the close relationship between humans and animals is a matter coming into more interesting question in our own time).

The way this conservation rule functions is clear. Animals can be eaten—but only within limits. As it turns out, a sacred territory is precisely the animal's breeding and water-hole territory. In times of severe drought or in breeding seasons, the animal must be spared, even in extreme conditions, to the point that even some of the tribe must be sacrificed for the sake of the animal. In this way a balance of human/animal populations is insured, in turn insuring a future.

A second, similar rule, again grounded in totemistic beliefs, relates to the rare unused or unusable animal. Leeches, for example, are not to be unnecessarily killed. And the reason given is that leeches belong to the great series of relationships that characterize the Aboriginal world: they are part of a close-linked ecosystem. Such an understanding has deep functional significance in a delicate desert environment. It is cast in prescientific terms but is "contemporary" with respect to sound ecological understanding.[3]

The point here, however, is not to recommend substituting Aboriginal culture for ours nor to simply romanticize the minimalist mode of life exemplified by that culture; rather, it is to make a contrast with our culture with respect to the relative role played by technologies. It is not simply the absence of complex technologies that makes the two cultures different but what could be called the two radically different ways in which the world is revealed. The Aboriginal, too, follows a dream of totalization, but a different dream—that of the Dreamtime.

The Dreamtime is "nature" brought into "culture" in a way different from the Heideggerian understanding of a resource well. Yet all of nature is "taken into account" in the Dreamtime, through stories that make the features of the landscape into sacred sites. Thus, the promontory is "where Goanna dreams" in a long and complex tale that is told and that also exemplifies some moral and tribal value. The water paths are "where the great snake wandered" in some other set of similar tales. The Dreamtime is a nontechnological taking of the world of nature into this culture, yet it is an attempted totalization, a mode of world revealed.

Functionally, such a mode of revealing a world, particularly when considered over the 40,000-plus-year history of these peoples, did little to modify the actual environment. And it was a cultural history which did not develop the fascination for artifactual invention shown in our own history.

So, at the end of this first itinerary, at the least, there is both a sameness to our trajectories in dreaming of totality and a deep difference. The dream of taking nature into culture technologically is shown to belong to a history, to *our* history. The enigma that appears at this junction is that our very distinctive history seems a currently dominant one. Aboriginal culture today is vestigial in the sense that, with isolated exceptions, the old ways are no longer being transmitted, at least not in their original forms. Even the vibrant religio-art of the peoples must now fit into a context which re-situates it into a crosscultural exchange, to be considered as an art commodity no longer belonging to its dreamtime relationship. But that is the fate of virtually all traditional cultures of this type.

Is there, then, a single, massive trajectory to the rise of high-technological culture and its attainment as a world culture? If so, the Marcuses, Jonases, and Elluls would turn out to be the prophets for our times.

References

Bennett, David H. 1985. "Inter-Species Ethics: A Brief Aboriginal and Non-Aboriginal Comparison." Discussion Papers in Environmental Philosophy, no. 7, Australian National University, Canberra.

Burke, James. 1978. *Connections.* Boston: Little, Brown and Co.

Descartes, René. 1953. *A Discourse on Method.* Translated by John Veitch. London: Dent.

Eisenstein, Sergei. 1949. *Film Form: Essays in Film Theory.* Edited and translated by Jay Leyda. New York: Harcourt, Brace and World.

Hacking, Ian. 1983. *Representing and Intervening*. Cambridge: Cambridge University Press.

Heelan, Patrick. 1983. *Space Perception and the Philosophy of Science*. Berkeley: University of California Press.

Hughes, J. Donald. 1975. *Ecology in Ancient Civilizations*. Albuquerque: University of New Mexico Press.

Ihde, Don. 1979. *Technics and Praxis*. Dordrecht: Reidel.

Levinson, Paul. 1985. "Toy, Mirror and Art: The Metamorphosis of Technological Culture." In *Philosophy, Technology and Human Affairs*, edited by Larry Hickman. College Station, TX: Ibis Press.

Mournier, Emmanuel. 1951. *Be Not Afraid*. Translated by Cynthia Rowland. London: Rockcliffe.

Whitehead, Alfred North. 1963. *Science and the Modern World*. New York: New American Library.

Chapter 7

Auditory Technologies

In order to continue to lecture, travel, and teach in my seventies, I now must wear a pair of hearing aids, acoustic technologies. Mine are small, in-channel, and state-of-the-art digital devices. These have three programs: one for everyday use; one tweaked to the "cocktail party" or restaurant setting in which near, ambient noise threatens to overwhelm hearing one's conversational partner(s); and one for telephone use. And, they are *expensive*!

There may be some personal irony here, more than three decades after I was engaged in all sorts of auditory experiments that became the basis for *Listening and Voice* in its first publication. Then, I had acute hearing and definitely, I did not need an acoustic technology as prosthesis. But, from then to now is a long time, and I propose to follow a phenomeno-logical itinerary along that way and analyze the process and experience of "embodying" such devices. I have described this process of embodiment in a series of previous works, beginning with *Technics and Praxis* (1979) and more definitively in *Technology and the Lifeworld* (1990): When we humans use technologies, both what the technology "is" or may be, and we, as users undergo an embodying process—we invent our technologies, but, in use, they "re-invent" us as well.

In both books mentioned, I used optical examples beginning with eyeglasses. Embodying eyeglasses, I contend, is much easier and somewhat different than embodying hearing devices. Again, I begin autobiographically—but any optician would recognize the implied patterns as symptomatic rather than individual: By my late fifties, I began to notice that it had become difficult to read telephone directories and the *New York Times*. So, after the proper examinations—themselves entailing sophisticated optical devices—I was given a diagnosis and a prescription for reading glasses. These I still wear, but I do not now need eyeglasses for nonreading purposes. However, there was a temporary time when I did need such mediating devices due to two closely timed accidents: chasing a porcupine in the dark after I heard him gnawing on my Vermont cabin, I encountered a tree twig and scratched my cornea. Not long after, one of my then infant children, with a finger poke, scratched the other cornea. So, after medical treatment, and then an eye exam, I had to wear prescription glasses for about a year while my corneas gradually returned to their proper shapes. I had to learn to see through glasses, to embody them.

If I revert to a third-person, anonymous description, I might recognize that what glasses do is to "correct" vision, in this case, to compensate for deformed corneal shapes. But just putting on glasses does not simply "snap" vision of the world into its now simply corrected sighting. Instead, one has to "learn" and bodily accommodate to wearing glasses. Phenomenologically, seeing is a whole-body experience. There are discernible changes in depth and motile perceptions, one is aware of this in the simple act of walking. The same happens with every new prescription. It is my bodily orientation that is the noematic part of this experience. This is even more noticeable if one wears reading glasses without taking them off to go get a drink of water—these mediating technologies produce a repeatable distorting effect that is quite perceivable. But, in my experience, and in those who have related theirs to me as well, embodying new eyeglasses to the point where they are nearly functionally "invisible" is a very quick process—taking maybe a day or two at most.

This is what I have previously called an *embodiment relation* with, in this case, optical technologies. I relate to my environment, my "world," by means of such technologies and if they are well functioning then experientially they are "taken into my very sense of bodily experience." My awareness of wearing glasses is a fringe awareness that gets interrupted only when there is back glare, or when the glasses slip off my nose, or when the lenses get dirty and smudged, when, in other words, something diminishes the normative transparency of the optics. "Breakdown" is a well-known phenomenon, made popular in the famous example of Heidegger's hammer—his claim was that only when something is missing or broke does the set in assignments and involvements become clear.[1]

With hearing aids, however, the technology of interest is an acoustic or auditory technology, a hearing "aid" ideally should function parallel to the visual eyeglasses example. Unfortunately, auditory transparency is much more difficult to attain, a fact well recognized by audiologists and others. A significant number of people attempt to use hearing aids, but the difficulty of embodiment is sometimes such that they give up.

I am not sure when I first became consciously aware of my slow loss of hearing. As with reading glasses, I became aware, with aging, that my hearing was not as keen as it once was. In academic life, verbal situations are of focal importance and at some point I began to be aware that the "cocktail party" hearing problem began to occur. That is, in a conference or reception setting, background conversations and other noise, seemed to intrude and overwhelm my ability to hear what nearer conversants were saying. Similarly, in large lecture classes, the questions from the back of the auditorium seemed too faint or indistinct. Beginning to recognize that I was experiencing hearing loss, and remembering all those sixteen-hour days of the loud noise of the two-cylinder John Deere tractor I used on my father's farm, I wondered if I had acquired "boilermaker's disease." However, I had also already noticed that many of my age peers among the faculty already wore hearing aids and their possibly loudest boyhood experiences were with stickball games. And, I knew I was too early to have the problem of Rock Concert disease either. But the clincher came via a more technological means.

While at an American Philosophical Association meeting in Boston, my wife and son were off visiting the Boston Science Museum and later invited me to revisit the place. One of the exhibits had to do with the senses, including hearing, and one could put on earphones and turn a dial to find out how many cycles per second one could perceive. I put them on and discovered that the upper range of my hearing only went up to 10,600 per second! Cognizant from descriptions of "normal" hearing that humans can hear between 200–20,000 cycles per second, I was shocked. I didn't say anything, but on reaching home, quickly went to my *Macropedia* to find out what the "objective" situation was: I was relieved, in one respect, to discover that my range was relatively "normal" for someone my age (early sixties then). What counts as normal, apparently, is also age-related. Recently I read a newspaper article in which a store owner, constantly having teenagers hang out in front of his store, wanted to prevent this pattern of behavior. Somehow he knew that the hearing capacities of those below twenty differed by a small frequency range from those in their thirties. He installed a noise device that broadcast this upper end of the frequencies heard only by those at the lower age level, but remained undetectable even to those in their thirties![2]

At this point, I decided to experiment. I purchased a single hearing aid, advertised as digital but not requiring a special set of exams. It did work for a short period of time—it amplified to the degree that I could better hear the questions in the back of the room and in low background noise situations. But its limitations were equally obvious.

Unfortunately, with more age, I experienced more hearing loss, now quite perceptible, particularly in the conversational contexts already mentioned, but also in a home setting. So, after a series of audiological tests that showed the degree of hearing loss I ended up with my pair of digital hearing aids. Undergoing such tests, again with complex technologies tuned to various auditory phenomena, I began to learn things I never knew before. Part of what I learned was that even high-tech, digital aids cannot restore frequencies lost, although within those that remain, these devices can be selective regarding enhancement or reduction and other manipulations. Again, this reflects my earlier claims about technologies—each transformation of experience displays an amplification/reduction structure. Eyeglasses do this; and so do hearing devices. But, in the case of speech a more subtle phenomenon arises. Vowels are temporally "longer" than consonants (and thus are more easily amplified); but since speech depends on the patterned gestalt of both vowels and consonants, the loss of consonants complicates hearing and understanding speech patterns. Digital devices, within limits, can enhance consonants. And, I experienced this with my first set of prescribed hearing aids. These were state-of-the-art devices, and they did make it easily possible to continue seminars, to allow much better auditory recognition of what the participants were saying. But, as technologically sophisticated as these were, once one was in the "cocktail party" situation, near conversants continued to be overwhelmed by the amplification of ambient surrounding sound.

If I refer back to *Listening and Voice*, it will be recalled that sound is simultaneously experienced as both surrounding and directional. Hearing aids, however, cannot simply match this phenomenon. And that was particularly the case with my first set. I had been advised to get a pair, and not a single device, even though I had more loss of hearing in one ear than the other. The reason was that I needed to relearn to hear directionally and this was presumably better accomplished if one began from the start with a pair of devices. Recall a parallel from optical history: monocles were once used, but are rarely, if ever, seen today. Instead eyeglasses as a "pair" are worn, although one lens may have a different grinding prescription than the other.

At first, I have to admit, while I recognized the improvement my devices provided, particularly in the conversation settings of home and seminar, the overall experience of hearing was clearly not anything like optical transparency with eyeglasses. My audiologist confirmed that this was, in fact, the normal experience for first-time users and urged me not to constantly remove the aids when I was not in the situations where they functioned best. He used the now-popular "the brain must relearn the process, so it needs the constant use to do this," also translated into, it takes a long time to become accustomed to hearing aids. My acoustic technologies remained quasi-opaque, although the recognition of this opacity was phenomenologically complex. For example, if I listened to music, if I didn't know the piece, I could not be aware of what I was *not* hearing (for example, the higher frequencies), but, if I listened to a piece that was familiar, I could quite distinctly become aware that I was missing what I remembered hearing in my listening past! Thus familiar pieces sound "odd," compared to how they "used to" sound. Other noticeable differences apply to "tinny" sounds in some of the retained sounds, particularly when coming from a stereo in another room. The phenomenon of what is lost, compared to that simply unheard, was evidenced in a seminar that Judy Lochhead and I sometimes share on the phenomenology of music. One evening, we were listening to a CD of sounds recorded by the ethnologist, Stephen Feld. He had recorded the very subtle sounds of water dripping, running, and so on, in the New Guinean highlands and was making a case for the auditory-dominant language of this tribe.[3] But when we played the CD, it was "silent" for me for the first minute or two—yet all the others were obviously hearing something, so, only indirectly did I become aware that I was missing something. (Chronologically, this event preceded my use of hearing aids.)

A counterphenomenon also occurs. In *Listening and Voice* I made reference to the fact that we always hear ourselves differently from the way others hear us. Physiologically, this is because we hear via bone conduction as well as through the sound waves carried environmentally. So, the "acoustic mirror" of the tape recorder is always a surprise at first. But, hearing aids *amplify* bone conduction sounds—thus my self-perceived voice is much *louder* to me with hearing aids than without. This complicates relearning of auditory projection when speaking, and even close-up conversations. My voice to myself seems "too loud" and so I try to compensate and end up having others tell me I am speaking too softly—or, sometimes, too loudly. This is the inverse of what is foreground and background in nonacoustic technology hearing.

All of this continued to be noticeable with my first set of devices; but as I approached the second anniversary of use, I lost the pair while flying from a conference in California back to New York (I take them out to listen to earphones if watching a movie or TV while in flight), and after not being able to find them or retrieve them via lost and found, I ended up getting a "new, improved" set. Indeed, in the two years of use, electronic and miniaturization technologies had already improved on the previous technologies.

The new set automatically "communicated" between the individual devices (if you change a setting on one, the other automatically picks up the change); there was a more refined set of programmable selectivities regarding frequency changes; and one or more programs could be made directional (two tiny microphones in each device enhance the sense of the directional, almost like having "four" ears!). All this for only $3,000 more than the original pair!

In certain ways, there were clearly perceptible improvements. Speech was indeed more distinct than with the previous aids; but most of all, in the second "cocktail party" setting, the directionality allowed me to hear close conversants even though the background noise remained amplified more than it would have without the devices. At least the cut-off point was better, but the amplification of background noise and the amplification of one's own voice still remains noticeable and frustrating. Technological "intentionality" is simply not the same as one's ordinary bodily intentionality. At the least, one can conclude that acoustic hearing devices are very far indeed from any of the hyped, utopian "bionic" beings of entertainment and science fiction imaginations or of virtual reality dreams. And, hearing aid embodiment does not come with either the same ease or degree of transparency that eyeglasses or optical technologies seem to have. Acoustic technologies are both more complex than the relatively simpler optical ones, but are still at a relatively early stage of development. But, I would contend, this is not because vision is in any way a "superior" sense, or a simpler one.

Hearing, however, is highly multidimensional—it implicates balance and motility in ways that implicate whole body experience intimately. Indeed, I would contend that technologies that come even closer to being prostheses for such more complicated experience, are more likely to be clumsy and less easily amenable to embodiment transparency. Just by way of one example, perhaps the oldest prostheses were "artificial" limbs, and while these have made various high-tech improvements, they have never yet allowed users to regain the gaits of walking, the gracious movement of dance, or more extremely with hands, the facilities of instrument playing that preprosthetic actual limbs allow.[4] Technological transparency, with respect to human embodiment, remains at best a quasi-transparency.

I shall not take up here alternative hearing devices, such as cochlear transplants, but I shall briefly look at some variations from other acoustic technologies. First, if one goes back in technological history, earlier "hearing aids" were often hearing horns or ear trumpets. One end small to fit into the ear, the other fluted out similar to the shape of a trumpet, such devices could mechanically channel sound and thus amplify it. In this case, the hearing horn, precisely because it enhances directionality and dampens ambient sound, did have a certain workability. Similarly, the stethoscope, which today remains a sort of icon of the medical profession, worked in a similar fashion. Indeed, the nineteenth century experimented with a whole series of auscultation devices, of which the stethoscope became the most successful one.

Auscultation, or the amplification of sounds from bodily interiors, when listened to by skilled and trained physicians, could detect heart murmurs, lung congestion, and a whole series of ailments that could not be detected either by visual or even tactile examination. Today, such skills are in decline and are largely displaced by newer diagnostic practices that rely on tests, or frequently, visual imaging (CT, MRI, PET scans, for example). When thought of as a variant on a hearing aid, the stethoscope, better than the hearing horn, dampens ambient sounds and amplifies sounds from bodily interiors. The tubes that carry the sound waves to the physician exclude ambient sounds in the examination room, and carry the amplified sounds of heart or lung processes instead.[5] Again, this is an example of a mediating technology displaying an amplification/reduction capacity. These acoustic technologies remain simply "mechanical." Today, most acoustic technologies of familiarity have become electronic, both analogue and digital.

Only briefly, these more contemporary variants include acoustic amplification capacities that go far beyond the mechanical display of acoustic phenomena. Technologies that include microphone-amplifier-speaker systems can make present sounds that cannot be heard at all without such technologies. Two years ago, I saw an interesting performance art example: in the installation, one lies down on a couchlike bed and listens to the amplified sounds of earthworms eating their way through a compost pile that is located directly below the couch. This is a less romantic, but in some ways biological equivalent to the amplification and recording of whale songs that are now part of familiar science documentary or enhanced music experience. With this style of instrumentation, however, one still hears frequencies within the normal range of hearing, thus the distant analogue to hearing aids is maintained while shifted to what could be called microsound. That is, the amplification of microsound is analogous to the optical microscope in that it makes the previously unheard hearable.

But perhaps the most ubiquitous acoustic technology that relates to hearing devices, is, of course the cell, or mobile phone, particularly those models that are worn in the ear. Here the variation is not directed at micro- or infrasound, but on the mediation of distance, which in its phone modality, is that of the voices of others. I have previously analyzed the irreal, near-distance that such communication devices display phenomenologically. Geographical or near/far distances are technologically transformed into the virtually same near-distance. Although I, myself, have resisted the habitual use of cell phones, even I have had the experience of calling my son to find out where he is, only to find out that he is home on the first floor while I am on the third floor in my study. His voice is as discernibly nongeographic and "near" as when the call is made from home to university. This now everyday experience of near-distance has become a familiar feature of the lifeworld.

The technological capacity to produce microsound and even to translate infrasound and to diminish geographic distance, may suggest that today's digital hearing devices may also eventually be able to follow such trajectories of development. Yet this is not to say that more radically constructed sounds could escape the constraints of all human embodiment. Rather, as with all prosthetic technologies, there will always remain "trade-offs." What are called trade-offs are precisely those interface clues to human-technology relations wherein we always remain short of "cyborgean" unity. I am sure that I am not unique or idiosyncratic if I admit that I would just as soon do without my hearing aids—just as those who wear glasses would like to do without those. But, the trade-off is precisely the strength of such prosthetic technologies and I am well aware that without my acoustic devices, I simply could not do what I now can continue to do.

Postscript

Since writing this chapter, I have acquired yet another, newer set of hearing devices—this time an open plan version that uses a transparent tube to conduct the sound, through a more open ear canal design. And, as with so much digital technology, once again marked, but incremental, improvements may be noted. The "stuffy ear" feeling of in-canal devices is gone; incremental improvement in suppressing ambient background noise occurs; and with a small remote control device, the adjustments are easier to make. In spite of these quite noticeable improvements, I remain far short of falling into the slippery slope belief that there will soon be anything like a fully transparent bionic recovery of "normal" human hearing. But the trade-off is better.

References

Feld, Steven. 1996. "Waterfalls of Song: An Acoustemology of Place Resounding in Bosavi, Papua New Guinea." In *Senses of Place*, edited by Steven Feld and Keith H. Basso, 91–135. Santa Fe, NM: School of American Research Press.

Ihde, Don. 1979. *Technics and Praxis*. Boston: Reidel.

———. 1990. *Technology and the Lifeworld*. Bloomington: Indiana University Press.

Kevles, Bettyann Holzmann. 1997. *Naked to the Bone: Medical Imaging in the Twentieth Century*. Reading, MA: Addison Wesley.

Olesen, Finn. 2006. "Technological Mediation and Embodied Health-Care Practices." In *Postphenomenology: A Critical Companion to Ihde*, ed. Evan Selinger, 231–45. Albany: State University of New York Press.

Sobchack, Vivian. 2004. *Carnal Thoughts: Embodiment and Moving Image Culture*. Berkeley: University of California Press.

Chapter 8

The Critique of Heidegger

A century after his birth, two very contrary statements can be made concerning Martin Heidegger: First, in a significant sense, he is surely one of the most important founders of the philosophy of technology. His insights into the structures and functions of technology remain deep and suggestive. Second, we all also know that he joined the National Socialist German Workers' Party and remained with it through the war. His associations with the movement, seen today as one of the most destructive applications of modern technology, are equally deeply disturbing.

My question is this: Is there something at the very heart of Heidegger's thought that makes both of these contraries possible? I begin my reflection with two vivid images, both related to that ancient Greek ancestry to which Heidegger turned again and again as a source of thinking, consonant with self-proclaimed origins for Euro-American civilization.

The first image is Heidegger's, that of the famous Greek temple in "The Origin of the Work of Art." Heidegger's temple is taken as a paradigm of artful *techné*, both "thingly" and signifying.

> Standing there, the [temple] rests on the rocky ground. This resting of the work draws up out of the rock the mystery of the rock's clumsy yet spontaneous support. Standing there, the building holds its ground against the storm raging above it and so first makes the storm itself manifest in its violence. The luster and gleam of the stone, though itself apparently glowing only by the grace of the sun, yet first brings to light the light of the day, the breadth of the sky, the darkness of the night.

The temple's firm towering makes visible the invisible space of the air. The steadfastness of the work contrasts with the surge of the surf, and its own repose brings out the raging of the sea. Tree and grass, eagle and bull, snake and cricket first enter into their distinctive shapes and thus come to appear as what they are. (Heidegger 1971, 42)

This "Wagnerian," this "Nietzschean" deployment of signification in the focal point of the Greek temple against the ground of its earth is typical Heidegger. It is heavy: it is *romantic*; it is what gathers the mortals, gods, earth, and sky. "The temple-work, standing there, opens up a world and at the same time sets this world back again on earth, which itself only thus emerges as native ground. . . . The temple in its standing there, first gives to things their look and to men their outlook on themselves" (Heidegger 1971, 42–43).

Heidegger's imagery is striking, captivating, and, above all, weighty. Now contrast what could be another look at the same image, done this time by a historian, J. Donald Hughes, in his book *Ecology in Ancient Civilizations*. "Those who look at the Parthenon, that incomparable symbol of the achievements of an ancient civilization, often do not see its wider setting. Behind the Acropolis, the bare dry mountains of Attica show their rocky bones against the blue Mediterranean sky, and the ruin of the finest temple built by the ancient Greeks is surrounded by the far vaster ruins of an environment which they desolated at the same time" (Hughes 1975, 1). Here, the same "thing," the Greek temple, reveals a very different "world" from that of Heidegger. Hughes goes on to point out: "In the centuries before the Golden Age of Athens, those same mountains were covered by forests and watered by springs and streams. The philosopher Plato saw evidence of the changes that had occurred not long before; there were buildings in Athens with beams fashioned from trees that had grown on hillsides which by his day were eroded and covered only with herbs, and he visited shrines once dedicated to the guardian spirits of flowing springs which had since dried up" (Hughes 1975, 1). What accounts for the dramatic difference in what is seen in the image of the Greek temple? Or, phrased even more starkly, is there, in the Heideggerian way of seeing, a deeper and even necessary way of *concealing* that allows only a romanticized perspective?

I opened this reflection with a well-known example from one of Heidegger's analyses of art objects. But his analyses of technological objects follow a close and similar pattern. Indeed, one could easily conclude that an art object is, for Heidegger, the primary example of a "good" technology. Both art objects and technological objects—equipment—are "thingly," "produced," have ways of "revealing" a world, and belong in some way to the process called *techné*, which Heidegger defines in the following passage: "There was a time when it was not technology alone that bore the name *techne'*. Once that revealing which brings forth truth into the splendor of radiant appearance was also called *techne'*. Once there was time when the bringing-forth of the true into the beautiful was called *techné*. The *poiésis* of the fine arts was also called *techné*" (Heidegger 1977, 315).

One cannot but detect, again, the heavy romantic overtones of this nostalgic merging of art and technology. Nor should we ignore from the outset that Heidegger's primary suggestion of a solution to the dilemmas of the Age of Technology often revolves around a kind of saving power found in art. But if this is where the revealing power that could save us from the reductions of modern technology is, it is because art and technology are closely related in precisely the thingly, produced, but revealing roles which both art objects and equipment or technological objects contain when they are seen as focal elements against a context or field that is "lighted up" as a "world."

Yet, in the Heideggerian corpus, there is often a great difference of evaluation and connotation between art objects and technological objects. On the surface, it might appear that the two most frequently patterned such differences relate to a certain suspicion concerning *modern* technology versus traditional technologies, and the older, smaller and simpler technologies versus the newer, larger, and more complex technologies.

There is much in the Heideggerian choice of "good" and "bad" connotations that commentators have noticed. Heidegger "likes" the tools of the workshop, the peasant shoes of the Van Gogh painting, the watermill on the stream, the windmill, and the old stone bridge with its arches. He does not like hydroelectric dams on the Rhine River, the atomic bomb, even the modern steel bridge that routes traffic to the same city square as does the old stone bridge. Such a pattern would seem to evidence a simple and old-fashioned romanticism of a nostalgic sort—and I would not deny that such a strain may be a found in Heidegger. But the issue is more complex than that.

There are, for example, inconsistencies: it is not always the small and relatively simple that escapes the Heideggerian disapprobation. Clearly, by today's standards the typewriter would be an example of a good, simply mechanical writing device, not much different in kind or principle than the Bavarian clocks beloved by the folk of Todtnauberg. Yet the typewriter receives a particularly scornful disapproving note:

> It is not by chance that modern man writes "with" the typewriter and "dictates"—the same word as "to invent creatively"—"into" the machine. This "history" of the kinds of writing is at the same time one of the major reasons for the increasing destruction of the word. The word no longer passes through the hand as it writes and acts authentically but through the mechanized pressure of the hand. The typewriter snatches script from the essential realm of the hand—and this means the hand is removed from the essential realm of the word.[1]

I virtually feel the scorn that would have been poured upon my composition of this paper with a word processor! But the point is that here is a relatively simple, mechanical device that does not escape the romantic thesis. I shall return to this example.

There are also much deeper inconsistencies in this pattern of choices of "good" and "bad" technologies. In the "Question concerning Technology," a deep danger of modern technology is laid to the way the world is revealed in the ensemble of modern technology as "standing-reserve," the extant translation of *Bestand*. (I translate this term as "resource well.") It is the whole of nature that is revealed as a resource well in modern technology, as illustrated through the disliked hydroelectric plant on the Rhine:

> The hydroelectric plant is set into the current of the Rhine. It sets the Rhine to supplying its hydraulic pressure, which then sets the turbines turning. This turning sets those machines in motion whose thrust sets going the electric current for which the long-distance power status and its network of cables are set up to dispatch electricity. In the context of the interlocking processes pertaining to the orderly disposition of electrical energy, even the Rhine itself appears to be something at our command. (Heidegger 1977, 297)

That is, nature, including the Rhine, is revealed as resource well, standing-reserve, for man's use. And while this production of energy does contrast with the old windmill—which can turn only when the wind blows and thus seemingly lets the wind "be"—it does not, in principle, differ from the smaller dam on the stream that allows the waterwheel in turn to grind the peasant's wheat. To allow this example as a "good" technology does not, to my mind, prevent seeing nature as resource well except in its lack of a larger interconnectedness with the electrical grid.

To this point, it should be clear that the romantic thesis, as I shall call it, pervades Heidegger's choices of "good" and "bad" technologies. But in what does it consist?

The first element, I claim, is a preference for what I call *embodiment relations*. Heidegger prefers, likes, those technologies that express straight-forward bodily, perceptual relations with the environment. This is part of what underlies his dislike of the typewriter. Expressivity, the connection with "word," is primitively found in a handy gesture: "Human beings 'act' through the hand; for the hand is, like the word, a distinguishing characteristic of humans. Only a being, such as the human, that 'has' the word (*mythos, logos*) can and must 'have hands.' . . . The hand becomes present as hand only where there is disclosure and concealment. . . . The hand has only emerged from and with the word" (Heidegger 1982, 118–19). As we saw with the typewriter, for Heidegger somehow there is less "hand" in writing with a typewriter than presumably that which is "handwritten" with a pen.

This same preference for simple, embodiment relations is exemplified in his very earliest analysis of human-technology relations, the famous "hammer example" in *Being and Time*. That example is simultaneously one of the most pointed in showing not only Heidegger's radical insights into technology but also a certain blindness and prejudice concerning technologies that do not express embodiment relations.

I shall not here go into great detail into this often-analyzed example, but shall instead note only a few salient features that are relevant to the way in which the romantic thesis also conceals important aspects of technology.

Positively, Heidegger shows in the hammer example that technologies in use are not objects as such; they "withdraw" in use and become partially transparent means by which humans relate to an environment. Here is a good critique of any simplistic and objectivistic account of technologies as simple objects. Rather, technologies are contextual, or field involved; the hammer "is" what it is in reference to the context of nails, project, and so on. It belongs to a reference system that always includes more than a mere hammer. Thus, while the hammer is always "thingly," it is never a *mere* thing and is, in use, transformed into a world-related and world-revealing way in which humans are involved with their environments. All of this—and more—is the source of the Heideggerian suggestivity for philosophy of technology.

But there is also a negative side to the analysis. In *Being and Time*, the context is "lit up" through technological *breakdown*. It is when the hammer is broken or missing that its involvements are shown. The fullness of the project—and the objectness of the hammer—gets shown when it is *not* functioning. I claim that here lies an early clue to a certain negativity that pervades the Heideggerian corpus and that blinds the analysis both to a possible appreciation of human-technology relations other than embodiment ones and to the features that, in fact, unite modern technologies to traditional ones. In *Being and Time*, it is hard to conceive of a positive relation *to* a piece of equipment, a technology, other than as that *through which* Dasein experiences its environment either in embodiment or with transparent referentiality. (The old turn-signal example from the German automobiles, which had flip-up arrows, manually operated, is what I call a *hermeneutic relation*, since actions are "read" *through* the technology. These are recognized in *Being and Time* but are on a close continuum with embodiment relations and are directly expressible in the technologically mediated action of the driver.)

In short, to relate *to* a technology in a positive way and in a situation in which the artifact takes on what I call an *alterity* relation seems to me inconceivable in the Heideggerian scheme. And although I cannot long belabor this here, an example from technology-as-toy may illustrate what I have in mind.

The child's top is just such a technology-as-toy that may become an *alterity relation*. Set in motion, the technology itself becomes an object of fascination. It has a quasi-life of its own, even apparent self-movement that is unpredictable. It becomes a quasi-other to which the child can happily relate. Such playful technological moments do not seem to belong to the heavy romanticism of the Heideggerian context. But just for that reason one could also miss the kind of appreciation and fascination that characterize much of the experience of modern technologies.

The preference for embodiment relations over other human-technology relations is what could be called a nostalgic element in the romantic thesis. It is hardly unique to Heidegger. It is also to be found in Karl Marx. Insofar as alienation theory is bound to any nostalgic element relating to the handwork of the worker prior to machine tools in a factory context, there may be found the same taste preference in that older mode of analysis.

A second element of the romantic thesis is one that has made Heidegger so appealing to those close to the environmental movement, particularly those now associated with "deep ecology." Here, too, a certain closeness between "good" technologies and the art object emerges more clearly.

A "good" technology is one that "gathers" a world in a certain way and "lets be" the nature and community that is so gathered. The old stone bridge in the essay "Building, Dwelling, Thinking," is an excellent example of this subthesis. Like the temple, the bridge reveals a certain world:

The bridge swings over the stream "with ease and power." It does not just connect banks that are already there. The banks emerge as banks only as the bridge crosses the stream. The bridge designedly causes them to lie across from each other. . . . It brings stream and bank and land into each other's neighborhood. The bridge *gathers* the earth as landscape around the stream. [And in a direct echo of the temple, the] . . . waters may wander on quiet and gay, the sky's floods from storm or that may shoot past the piers in torrential waves—the bridge is ready for the sky's weather and its fickle nature. . . . The bridge lets the stream run its course and at the same time grants their way to mortals so that they may come and go from shore to shore. . . . The bridge *gathers* to itself in *its own way* earth and sky, divinities and mortals. (Heidegger 1971, 152–53)

Here art object and use object merge positively. The bridge is clearly a "good" technology when it gathers the Heideggerian fourfold in its focal/field relation. Yet, not all bridges do this, according to Heidegger: "The highway bridge is tied into the network of long-distance traffic, paced as calculated for maximum yield. [This stands in contrast to the] . . . old stone bridge's humble brook crossing [that] gives to the harvest wagon its passage from the fields into the village and carries the lumber cart from the field path to the road" (Heidegger 1971, 152).

The steel highway bridge is to the old stone bridge precisely what the typewriter is to the pen. Yet even in the midst of this clearly romanticized difference, Heidegger must admit that both bridges have something in common: "Always and ever differently the bridge escorts the lingering and hastening ways of men to and fro, so that they may get to other banks and in the end, as mortals, to the other side" (Heidegger 1971, 152–53). This means more than that a bridge functionally connects the two sides of the river. It means that a bridge, ancient stone or modern steel, *gathers* in its way. Authentically or "good" sounds in the term "lingering," whereas unauthentic or "bad" in the "hastening," which belongs to the highway network. So, gods and mortals, sky and earth are gathered—although differently—by both bridges: "The bridge *gathers* as a passage that crosses, before the divinities—whether we explicitly think of, and visibly *give thanks for*, their presence, as in the figure of the saint of the bridge, or whether that divine presence is obstructed or even pushed wholly aside" (Heidegger 1971, 153).

Gathering, the deep signifying surplus of meaning in the figure/ground of bridge and landscape, thus belongs to any such artifact. The difference lies in Heidegger's late variant upon how this may happen authentically or unauthentically. And while the romantic thesis clearly belongs intimately to this distinction, it does so in a complicated way.

The authentic mode of gathering is one that states, "To preserve the fourfold, to save the earth, to receive the sky, to await the divinities, and to escort mortals—this fourfold preserving is the simple nature, the presencing, or dwelling" (Heidegger 1971, 158–59). It is at least implicitly clear that the conservative view of Heidegger's notion of Germanic life is what best fulfills this preservation. That, too, is part of the romantic thesis. The authentic mode of gathering contrasts with its unauthentic counterpart, the world as revealed through the "gathering"—which inauthentically becomes *enframing* in the worldview of modern technology.

I am quite aware that Heidegger does not simply outright condemn modern technology—its essence, enframing, is simultaneously a revealing of the world and an openness: "When we consider the essence of technology we experience enframing as a destining of revealing. In this way we are already sojourning within the open space of destining . . . [and] when we once open ourselves expressly to the *essence* of technology we find ourselves unexpectedly taken into a freeing claim" (Heidegger 1977, 307).

But there is danger here of the inauthentic closure and reduction that arise from this form of gathering: "The coming to presence of technology threatens revealing, threatens it with the possibility that all revealing will be consumed in ordering and that everything will present itself only in the unconcealedness of standing-reserve" (Heidegger 1977, 315). In short, all of nature, including human being, will be seen as reduced to a vast resource well—but the question then is: for whom, or for what end?

So, the stone bridge or the steel one, the temple or—and here I shall introduce a counterexample bound to be outrageous to Heideggerians—the nuclear plant gather the fourfold, albeit in different ways. Here, then, is my post-Heideggerian example:

> Seen while sailing in Long Island Sound, on the horizon stands the stark super-silo, light green topped, of the Shoreham nuclear plant. Standing there, it brings to presence the very contrast between the seemingly featureless sandhill earth with the sky. It stands at and defines the contrast, too, between the sea and the

shore, which without its focal presence would also be featureless lines along the horizon.

I could, of course, go on in this crypto-Heideggerian mode, but the point is made. But if the nuclear plant substitution for the temple is somehow outrageous, what makes it so? I contend that the difference is not simply the difference between the nostalgic romanticism of the Greek temple and the urgent and fearful presence of the nuclear plant. Rather, it lies in what is left out, concealed, or unsaid in the Heideggerian account.

What is left out of both the Heideggerian account of the temple—and of the crypto-Heideggerian paraphrase concerning the nuclear plant—is what Langdon Winner has called the "politics of the artifact." For us, that dimension of the thingly is more vividly present in the nuclear plant than in the lost civilization of the Greeks only because it is nearer to us.

We know that the Shoreham nuclear plant was sold to New York State for one dollar (its cost had been five billion dollars). We know that the reasons had to do with the recognition that Long Island, the most populous nonurban area in the United States, could not be evacuated in a Three Mile Island or Chernobyl-type accident, and with the political opposition of over 75 percent of the populace to its going online.

Yet, ironically, precisely because the nuclear plant *revealed* its artifactual politics in a somewhat Heideggerian fashion, this first stoppage of a plant before opening could occur. It was because the Shoreham plant revealed—negatively—its form of gathering that it could be closed. So, as at the beginning of this reflection where I juxtaposed J. Donald Hughes's reading of the Greek temple with Heidegger's, the same can be done here. In its gathering, the nuclear plant makes the fishy life of the Sound to appear—as drawn to the warmer exhaust waters of the plant, but to be placed in danger of a leak, as in the case of the Irish Sea, which today is the world's most radioactive sea because of the irreparable leak at Sellafield. It channels the community into its pathways, now recognized to be more cloggable than the Long Island Expressway at high traffic time. It reveals the hastening which would be needed to evacuate its wastes (by sea, or by land?), and so on. And I add as a postscript to this Hughesian variant upon Heidegger that I could not borrow the even more poignant image of Langdon Winner in his description of the Diablo Canyon plant, beyond which he sighted a gray whale in the Pacific. This is because Long Island Sound has ceased to entertain whales or porpoises for the entirety of the twentieth century and, as of 1987, has its first forty miles of bottom so hypoxic (absent of oxygen) because of excessive phosphate and nitrogen runoff from modern sewerage that even the lobsters are dead in that area.

By adding this politics of our artifacts to the analysis—absent from the Heideggerian account, concealed and unsaid—the account becomes even more powerful. What needs to be noted, however, is that the romantic thesis in its unsaying concealment has all along hidden this politics of the thingly. It hid the Greek politics of the thingly just as well as it hides ours.

Hughes, whose image of a temple surrounded by a decimated environment, points out that every ancient civilization of the Mediterranean Basin brought about the same result: the Greeks deforested Attica; the Phoenicians, Lebanon (whose cedars went to Israel to build temples not unlike the Athenians); the Latins, the Italian peninsula—and all with low-technology bronze axes.

We, of course, recognize some difference between this ancient rise of civilizations and their impact upon the earth and the modern equivalent. Brazil's rapid decimation of the rainforests is speedier because of chain saws and tree-cutting megamachines, hypocritically echoed in our own free cutting of the redwood forests of the Pacific Coast. High technologies amplify and magnify what a culture can do; and today we are aware of the threat of the greenhouse effect, whose gases, according to a very recent account in *Science*, are probably 20 to 25 percent homogenic in origin.

But now I have gone too far. My illustrative excesses may make it sound as if Heidegger was not one of the most important "fathers" of contemporary philosophy of technology and, worse, that those of us who are post-Heideggerian are even more negative on modern technology. Neither of these conclusions is my intent. Thus I must make two closing apologies.

To make my apologia to the more reverent Heideggerians among us, I recognize that one of the deep significances of his account of technology is to have shown it not to be simply some collection of objects, but long pre-Kuhnian, a way of seeing, of revealing a world. And Heidegger is right, at least about the dominant way of Western seeing of nature as a resource well for human purposes. What is uniquely Western about this view lies mostly in its connection to the systematic mathematization of nature and the emptying of nature of any but inanimate and deanthropomorphized qualities. Unfortunately, many—indeed, most—prior cultures have also included components that frequently negatively affect environments.

A survey of so-called primitive cultures would reveal a startling ability—with very low technologies—to devastate environments. Slash-and-burn agriculture could subsist only in a rich, renewable forest so long as the human population remained low. The Pacific Islanders, who swept eastward centuries before Leif Eriksson got to the New World, extirpated species after species of wingless and winged birds in the easy search for food and fashion. And one could go on gloomily for some time. The exceptions of finely balanced minimalist societies such as the Inuit in the Arctic or the inland Aboriginals (for forty thousand years) in Central Australia, who left environments little affected, are also rare in previous times.

So, now my second apologia: it is not simply modern technology that threatens the environment. It is only the extent and the amplification of power that make it global rather than regional. My negativity, rather, is addressed at cultures that embed technologies but heretofore have too often not been sensitive to the contextual and long-range effects of technologically enhanced human action.

Here, in conclusion, begins to emerge my reason for so thoroughly demythologizing romanticisms: there is no previous time to which we can return where the gathering of the fourfold was "right." The Greeks, the Romans, the Hebrews, our forefathers were not sufficiently concerned with our earth in any of their forms; neither were the peasants, who for all their preservation and building up of soils in Europe, were still the world's most populous per square mile on the earth. They had the politics that allowed the same peasants who preserved the soil to harbor the most virulent anti-Semitism and nostalgic call for a purified homeland, which nurtured the Nazism of Heidegger's time. This ambiguity of preservation of the earth with destruction of a whole group of peoples is matched by another unnoted ambiguity in the Heideggerian corpus.

The networks—highways for the steel bridge, the electric grid for the hydroelectric plant—are also ambiguous. Because in the modern—and now I would say postmodern—world the network is what is beginning to make us aware of the displacement of our chauvinistic Eurocentrism, which, to my mind, is linked with the romantic thesis not only in Heidegger but also in our dominant views of technology, nature, and each other's cultures.

What is needed is not a rejection of the deep and essentially phenomenological insights into technology as a culturally embedded phenomenon with its different gestalt features, but a deepening and more complex appreciation of all of the facets of our technologically textured mode of life. That includes and must include the explicit recognition of both the politics of our artifacts, and the demythologization of nostalgic and romantic views of previous times.

Instead, we need to develop a postmodern critique that, at this early juncture, is still in a bricolage stage. But out of our growing experience of cross-culturality, we have begun to recognize that there is a plurality of cultures out there that threaten to decenter our past assumptions, and alongside which—but only alongside—we must reevaluate our past assumptions.

So, my demythologization of romanticism is also a critique. It is aimed at noting the freeing side of postmodern technological civilization and the opportunities that lie in its very networked ambiguity. Global pollution, the threat to the earth posed by our amplified powers, has also the promise of now seeing ourselves globally within a plurality of cultures. None of these should, or ought to be, romanticized. Rather, our emerging but still primitive awareness of pluriculture should be taken only as a threshold for simultaneously freeing ourselves of a past fraught with too frequently had ambiguities and opening ourselves to the uniqueness of a new world, equally ambiguous, but for the first time genuinely global.

The dramatic space shots of Earth from the moon or a satellite are very un-Heideggerian precisely because they place Earth at a distance from Earth-as-ground. But they are also irreversibly part of the postmodern view of Earth-as-globe, with a very different sense of what constitutes our "home."

References

Heidegger, Martin. 1971. *Poetry, Language, Thought*. Translated by Albert Hofstadter. New York: Harper and Row.

———. 1977. *Basic Writings*. Edited by David Farrell Krell. New York: Harper and Row.

———. 1982. *Parmenides*. Frankfurt: Klostermann.

Heim, Michael. 1987. *Electric Language*. New Haven, CT: Yale University Press.

Hughes, J. Donald. 1975. *Ecology in Ancient Civilizations*. Albuquerque: University of New Mexico Press.

Chapter 9

Multistability and Cultural Context

Philosophy of technology is a recent field within academic philosophy. Its roots, as both Rapp and Ihde have shown (Ihde 1979; Rapp 1981), are multiple as are its thematic questions. Historically, however, it remains the case that what is today called philosophy of technology is largely a Euro–North American discipline. It has risen at a time when it is almost impossible for critical and reflective thinkers *not* to thematize what has become a highly important factor in virtually every dimension of human life. Ours are technologically saturated cultures, and in that respect they are both qualitatively and quantitatively different from cultures which have preceded us. But the technology about which we are prone to reflect is both a scientific and industrial technology. Thus it is no accident that in the Euro–North American context there are concentrated groupings around two sets of "internal questions."

On the one side, there arises a set of internal questions related to the classical concerns of philosophy for epistemology, ontology and metaphysics. These include issues which are common to the philosophy of science and its history. For example, questions arise about how technology is related to, differs from, or has effects on science. Euro-American philosophers inquiring into such matters include a wide range of figures—from positivistic and analytic thinkers who usually emphasize a strong difference between scientific and prescientific technologies, and who often maintain an epistemological primacy to theory over practice, of science over technology. But phenomenologically oriented thinkers, most notably Heidegger, raise the same question while inverting it, making science the necessary "tool" of Technology which itself is taken as a mode of revealing a "world." In both cases, however, the very traditions of science are taken for granted.

On the other side, there arises a set of questions more relevant to ethics, social and political philosophy, which tend to focus more upon the social embodiment of contemporary technology, i.e., corporate or industrial technology. Here issues of alienation, exploitation, dehumanization, the bureaucratization or mechanization of life, take precedence. And again the spectrum of thinkers is at least as wide as those who, on the one side, adapt a neo-utilitarian stance to create an ethical calculus to deal with medical technologies, to the other side, Marxists and Critical Theorists who deal with communication, ideal languages, and questions of who benefits from technologies.

The development of such a discipline is to be applauded, particularly as it is the beginning of a critical reflection upon precisely one of the most obviously powerful of human impacts upon today's earth. And what I have called "internal questions" are necessary to address (as I, myself, have done; Ihde 1979, ch. 10). But there are two other types of questions which are also needed, if philosophy of technology is to become more comprehensive than it is to date.

Whole-Earth Measurements

One of these dimensions is what might be called the *global dimension*. There are many pressing questions facing us today which can only be answered globally, in terms of whole-earth measurements. For example, in earth science one such global question is whether or not the earth as a whole is cooling off or warming up? Is it headed for a new ice age? Or a greenhouse effect? The answer to this global question is of obvious import to us humans and our future. Yet, we are also aware that we do not have the means to make whole-earth measurements either with respect to spatial extent or with respect to long time records. (Although with satellite instruments we are beginning to approximate the spatial measurements needed.)

Similarly, and even more urgently, we have questions of global import which arise from the proliferation of nuclear weaponry and politically desta-bilized economies. Is a nuclear war more likely now than previously? Does the use of "tactical" nuclear weaponry necessarily lead to global holocaust? etc. Again, we have no whole-earth means of answering such questions although the answers are urgent.

I suggest that there has been a similar global question which has been a background question in the philosophy of technology—at least often dealt with by speculative thinkers who wish to deal with contemporary technology at a global level. Is there some overwhelming or even deterministic trajectory to the development of contemporary scientific technology? Both utopian and dystopian speculators have intimated that the answer is "yes." Utopians, perhaps less dominant now than in the nineteenth century, have predicted often enough that the application of scientific technology will very soon lead us into a world of leisure, the solution of social problems, and perhaps even the literal solution to everlasting life. Dystopians, contrarily, prophesy a kind of totalitarian closure made possible by the political application of a universal technology, or else the end of the world by a "bang" of nuclear war or the "whimper" of global pollution. Many more thinkers—take Marcuse and Ellul as two—speculate that even short of the above effects, scientific technology includes the probability of a universal culture, lacking in variety, and ultimately boring if not worse.

Both utopians and dystopians share some often unstated premises: that the use of technologies is in some degree deterministic; that technologies once invented and employed, exceed the abilities of humans to "control" them, a frankenstein effect; and that as a certain kind of technology becomes worldwide, it necessarily follows that its impact will likewise be one of reducing human culture to a technological uniformity. Both utopians and dystopians presume a Technology with a capital.

Unfortunately, as with the problems of whole-earth measurements, it is not at all clear that we have the concepts or tools which would be adequate to such an assessment. Yet the questions implied in the speculation are genuinely important and even urgent. The approach I shall take here attempts a preliminary and perhaps more modest approach to these questions.

Related to the global question is another: Given the obvious fact that Western scientific and industrial technologies are now spreading globally, is the trajectory one which increasingly points to a uniform development? Or, as technologies embed in different cultures, will there be multidimensioned results? Here we reach a question similar to the one Martin Heidegger raised in his famous "Question concerning Technology" (1977, 283–318). I shall position myself with respect to this paper in terms of both this philosophical tradition and a set of contemporary experiences.

First, the tradition: Heidegger's essay displays what I take to be a deep insight. Technology as a phenomenon can be better understood as a means of seeing a "world" than simply as a complex of tools used by subjects. Technologies reveal worlds. I take it that what is referred to in this insight is what I shall call cultural embeddedness—technologies are historico-culturally embedded and this embeddedness entails a whole view of a world, including such dimensions as space and time.

However, having adapted this Heideggerian view, I shall move away from the implicit claim of his essay. His question was: what is the essence of technology? It remains ambiguous as to whether he meant simply the essence of Western technology, whether that essence was simply a globally victorious one and hence *the* technology of our era, or whether he was addressing only our internal tradition. Whatever is the case, a recapturing of an older Husserlian technique can be readapted as a critique of the Heideggerian result. Husserl claimed that the way to establish or discover essences or invariants, was by the use of *variations*. I have increasingly come to the conclusion that variational method is, in fact, the very heart of a phenomenological inquiry, and I shall apply this method to the questions at hand. In this case the variations will be cultural ones, variations upon culturally embedded technologies and their entailments.

By drawing from history and anthropology, there appears a rich field of differing ways in which humans have embedded technologies and through them experienced worlds. Contrarily, I am offering an implicit critique of Heidegger in that if it is correct that invariants can only be demonstrated through variants, it is clear that he does not utilize variants and thus his claims concerning the essence of technology must be at least reduced in scope.

Now, the experiences: My own itinerary in philosophy of technology arose from the center of Euro–North American thinking. Moreover, what I took to be questions central to philosophy of technology arose from precisely the questions mentioned in the introduction. But within the last year and a half I have encountered situations which have made me radically rethink the context for philosophy of technology. I shall briefly recount two such occasions.

The first occasion was in Colombia, South America. In February 1982 I conducted an intensive faculty development seminar series on the philosophy of technology, which was attended by persons from some four major universities. What I had intended originally was to deal with precisely some of the questions from the "internal" set of questions mentioned, including the problem of the relation of science to technology. But by the second session it became dramatically clear to me that something was wrong with the situation. To put it briefly, I became aware that what we take for granted concerning science and science education, was simply missing in this situation. Indeed, the phenomenon "scientific technology" was there perceived simply as a unity, both "science" and its "technology" were perceived as politicocultural. Scientific technology—better scientific-industrial technology—was the expression of an alien culture in the wake of which impact could be seen threats to ways of life, cultural myths, and virtually everything relating to the Latin form of life. This is to say that the very perception or interpretation of the phenomenon was different. What I could take as an important question: the relationship of science to technology, was there seen as merely an internal and probably ultimately unimportant tinkering with an overwhelmingly powerful phenomenon which had other more immediate effects. I began to realize that just as the first-time visitor to a new land can perceive things which the native takes for granted or overlooks, so those who experience the coming of what we take for granted can perhaps see what we do not see. At the very least, whatever the past may have been, science-technology in its contemporary incarnation is a unitary phenomenon, and it is the expression of a particular culture and its world—that the Latin Americans saw more keenly than had I. Reflectively, I began to discern our own tendency to parochialism.

The second occasion was not entirely different. In this case the visit was to a group of African universities in the summer of 1982. I was particularly intrigued with what I found in the Black universities. Once again it was obvious that science-technology was simply taken as a unitary phenomenon, as an already accomplished symbiosis. Equally obvious was its appearance as the historical-cultural expression of Euro-America. And its impact was one which could be seen to impact primarily upon indigenous culture. Only in this case, contrary to the threat felt by the participants in Latin America, the impact was seen as the very way to attain a future. There was, in the Black universities I visited, a belief that by the virtual total adaptation of a technological culture, one could leap ahead into a twenty-first century culture which at the least was unlike the traditional cultures which for the students were increasingly perceived as archaic and of little interest. Indeed, the parallelism with second generation immigrant children in the United States came to mind. I think of my father's generation (second) in which there was a reflection of the old-fashioned ways of their fathers and a virtually total, almost religious belief in American progress and technology.

What emerges from both these experiences is this: Contemporary technology is perceived and interpreted differently in the Third World context. Its "internal" field, that is, its history from the Greco-European West, and its science, at least as a focal framework, is submerged. Neither are the Third Worlders aware of this context, nor ultimately interested. Instead, the perception of contemporary technology is of a cultural-political context, much like the invasion of a new people, and the response has to be one of (a) do we fight? (b) do we accept? or (c) do we adapt in part and resist in part? And in these Third World contexts, we find a rich field of possibilities precisely because of this bifurcation in what constitutes the context of the taken for granted.

Here, however, I must return to the problem raised with whole-earth measurement. How can we get hold of the cultural embeddedness of technologies? And, how can we gain some distance such that we can see both our own and variant forms of this phenomenon?

To answer this question, I propose to use a phenomenological technique drawn from variational theory to analyse three levels of culturally embedded technologies. If there is an essence or invariant to technologies, it is to be arrived at through variations, and in this case variations are historico-cultural ones with case histories drawn from historical and anthropological materials. The levels I shall examine will include (a) a high altitude look at some historical examples; (b) a comparative look at widely variant cultures with emphasis upon differently perceived "worlds"; and (c) then a return to the situation of technology and its interrelation between Euro-American and Third World cultures.

High Altitude and Technological Embeddedness

If we were to begin with a high altitude survey of the ways in which technologies are culturally embedded, it would not take much insight to recognize ours (Euro-American) as a *technologically saturated* culture. Technologies play a role in virtually every activity we undertake and are implicated in both the mundane and the most intimate of our daily affairs. Our very sexuality in increasingly mediated technologically, whether through mundane uses of birth control or through the life-determining techniques including artificial insemination, abortion, test-tube pregnancy transfers, or whatever. Indeed, I dare you to find a single ordinary activity which does not implicate and engage technologies both as individual artifacts and as implicated systems. Who today simply urinates in a field or wood—a most basic activity— rather we engage an entire waste and water system, an architecture, etc., all implicating a tie-in with our technological culture.

If ours were to be taken as one extreme end of a continuum in terms of saturation, we would find at the other end of the continuum only a few remaining "primitive" or "archaic" cultures, such as a few found in South America. Note that these societies are not without technologies. The system of hunting which utilizes nets, bow and arrow, even sophisticated poison systems, are technologically mediated ways of relating to an environment—but compared to ours, we could term this a *minimalist* technological culture.

My focus is upon the variant ways in which technologies are culturally embedded, and it is preliminarily important to note that what often is taken as a kind of linear determinism is not verified in either history or anthropology. For example, the generalization often made that certain kinds of technologies—presumably high and complex technologies—incline or even determine centralist social organizations, while simple and low technologies determine toward village or decentralized organization, does not seem to be the case.

As Lewis Mumford (1971) has shown, a society with relatively simple and low technologies may, nevertheless, be highly centrally organized and attain major feats of monumental building. The Egyptian pyramids and the Mayan cities are two such examples of historicocultural variants which utilize low, simple technologies embedded in high, centralized and hierarchical cultural-political organizations. Contrarily, in today's world, decentralist tendencies are often made possible by sophisticated, complex technologies. For example, centrally located and thus easily controlled media such as newspapers, have in countries which are under revolutionary pressures, often been replaced by the virtually uncontrollable dissemination of cassette and miniature tape recorders for the propagation of nonconformist news and propaganda.

This is not to say that some particular technologies lack specific inclinational *teloi*. Current types of nuclear power systems obviously are correlatable virtually only with centralized organizations, while solar systems may be either centralized or decentralized, and wood heating is likely to be best utilized in a decentralized context.

While what I am suggesting at this high altitude level is critical of some too-quick conclusions of speculative global thinkers, it is more a warning not to overgeneralize about either Technology with the capital, or about the way technologies may be culturally embedded. Rather, I suspect that the way technologies are culturally embedded as variants, relates in a more *gestalt* fashion to human forms of life. This direction is hinted at by much of Lynn White's work. Thus the same technology may be used in quite different ways in variant cultures. The wind driven rotor was used as a prayer wheel in India, but as a windmill for power in Europe; black powder was used for fireworks and celebration in ancient China, for gunpowder in the West. Here the same technologies embed differently in variant cultures.

The situation is clearly highly complex and were we to have a table of elements which could be correlated with the possibilities, the task would be suitable for our largest computers—yet if we are to suggest what the parameters for whole-earth measurement might be, we must open the door.

Cultural Variations

To do this, I shall now descend to mid-altitude and undertake a more precise analysis of some major variables in variant cultures. Clearly not all variables can be dealt with, so I shall select a few which are nevertheless highly suggestive with respect to the problems of cultural embeddedness. In typical phenomenological fashion, I will begin with perceptual dimensions. The factors I shall examine include:

1. *Perceptual origin or bodily position.* While bodily position is in a primary role with respect to first person perception, its role differs widely with respect to where it fits into our dimensions of technology and cultural embeddedness. Yet, in some primitive sense, bodily position is the point-zero from which perception takes place.

2. *Perspective or privileged point-of-view.* Related to perceptual origin, but not reducible to it, is the position from which a perspective may be taken. While there is usually a hypothetical position implied in all perspectives, it is not always equitable with bodily position.

3. *Space-time horizons.* Space-time horizons are particular interpretations of perceptual gestalts. Here I am taking a phenomenological "Einsteinian" stance which, in effect, does not conceive of space-time as a Newtonian empty and absolute container, but instead takes space-time as some configuration of material interrelations and thus for our purposes a type of perceptual gestalt.

I may illustrate how these variables work by drawing briefly from the example I elaborated at the last German-American philosophy of technology conference: Then I elaborated the vast difference between European and Polynesian systems of long distance spatial orientation, i.e., navigation. At that time I was primarily concerned to demonstrate that a technologically mediated (instrument utilization in a mathematical interpretation) and a perceptually read system varied. In that case I contended that Westerners, from the beginning, presumed a technological relation to the world, mathematically contexted, while Polynesians a perceptually direct, but hermeneutic or "read" relation to the world as means of long-range orientation.

In this context I can be even more specific with respect to the variables mentioned. First, I shall discuss perceptual origin and perspective. In the case of Micronesian navigation, as Thomas Gladwyn (1970) has so elegantly shown, bodily position and perspective are identical. Even I did not realize how radical such a position would be with respect to our taken-for-granted gestalt until recently. The navigator in this system quite literally takes his actual bodily position as the fixed variable for navigation. Thus even the interpretation of primary perception takes on a sense different from our usual intuitions.

The boat does not "move through the water" as we might say—which from our set of questions now may be seen to imply a perspective *other* than that which we actually occupy—but, rather, the island being sought moves toward the navigator, or, more immediately, the water moves past the boat as the ocean moves in relation to the boat.

The space-time horizon, then, is in motion with respect to the fixity of the navigator. Location is arrived at by noting such phenomena as "ghost islands" which are reference directions in this system of navigation that utilizes a kind of bodily geometry, but not a mathematical geometry.

Contrarily, we ordinarily think of ourselves as moving toward a point on a map, for in effect, we have separated bodily position from privileged perspective. Were we to occupy the perspective which we assume, we would be, for example, in a satellite above the ocean upon which we are moving, a "bird's-eye" perspective which indeed became the privileged perspective for our long-range orientation.

Our space-time horizon is interpreted from the fixed heavenly or bird's-eye position which is fixed while we move on the surface of the earth. This, in turn is interpreted as position on a grid of coordinates which imaginarily covers the surface of the earth. Note that the immediate perceptual phenomena, an example of which is the boat in the water, can make equal sense—I shall say multistable sense—in both cases. Here we have an "Einsteinian" relativity of perspectives: were the situation to be one of a canoe in a river with a current, it might be immobile with respect to the shore, moving with respect to the current, and could be said to be either moving against the stream or not moving with respect to the land. Both are correct, but relative to the point of variable from which the observation is made.

A. F. Aveni, an astronomer-anthropologist at Colgate University, has taken this same notion into the area of archeoastronomy which is the discipline which tries to understand and reconstruct ancient systems of astronomy. In a brilliant article, "Tropical Archeoastronomy," Aveni (1981) shows that observations of the heavens in tropical as contrasted with temperate regions are quite different. Star paths are, for instance, widely variant. In tropical regions stars traverse the heavens basically vertically from the horizon and set in the same fashion. This movement makes what Westerners call a "sidereal compass" a useful reference point (and this technique is deeply engrained in South Pacific navigation as well). Contrarily, in temperate regions which are north by virtue of the earth's land masses, star paths may be fixed (Polaris or the North Star), elliptical (above the poles), or vertical (near the equator). Thus the actual observable movements are different from the two different bodily positions of peoples observing the heavenly bodies. Aveni was able to show that there was a similarity of astronomical concepts throughout the tropical world which varied from those of the temperate world. Tropical systems use a "horizon" or sidereal system, while temperate systems employ a "polar" reference plane.

Tropical systems, for example, pay little attention to equatorial or ecliptic references, but all seem to have some version of cardinal directions from the horizon (sidereal compass), in fact with an amazing similarity of interpretation regarding pillars or ridge poles reaching into the sky (Aveni 1981, 163–64). This domestication of the horizon as a place of dwelling also reveals something about the familiarity with the earth and ocean included in archaic culture. A curiosity emerges when, in fact, our instruments are introduced into this scene. The Puluwati, for example, now use the compass as a steering device, substituting its ease for the finer observations needed to read wave patterns. But its circularity and the evenly divided segments posed no problems for this people who considered the earth to be square and the guide stars on the horizontal points unevenly distributed. This is because each of these elements are used for direction relative to the navigator, and neither for distance nor world-mapping—the result is equivalent or multistable regardless of whether the earth is round or square. Indeed, if the perspective and bodily position are identical, and you are situated at the same level as the horizon, the observation could be of a circle, a square or a polygon—only if you assume as outside or above or birds-eye perspective does this interpretation get called into question.

In ancient astronomy a similar observation may be made in that all heavenly bodies are frequently regarded to be equidistant from the earth (as holes in the heavenly dome, for example) or, if arranged in concentric circles of distance, usually so arranged in terms of some cultural set of values.

Two additional items are worth noting here: first, our tendency is naturally one which tries to bring what is radically different for us into our familiar world—thus even Aveni speaks of "star compasses" and the like, which in effect read our technologically embedded techniques into those of peoples who do not read things this way. And, second, we should note that contemporary, scientific astronomy does not in a fundamental sense belong to either of these archaic systems. In that regard it may be well to note that ours is, with respect to perceptual origin and perspective, thoroughly "Einsteinian" in that it is not even possible without a multiplicity of relative perspectives (from earth, from the sun, from our galaxy, and so forth.)

The contemporary astronomical system may be related to the three variables here as well. First, bodily position or origin of perception is at most "accidental" or arbitrary—it happens to be where I am. But even this is not often the case in a full technologized instrumental context. Much astronomy is done photographically, or through radio telescopy, frequently automated. Thus the actual activity of the observer is distinct from the instrumental use in many cases. Second, as already noted in Western navigation, the emergence of a disembodied or imaginary perspective, often privileged as more important than bodily position, is here taken one step further. Actual observations from satellites or imagined perspectives from some distant point in the universe are the norm for contemporary astronomy. Finally, even the space-time dimensions of such an astronomy become relativized and representation may take the shape of a two-dimensional representation of a "flattened" universe; a three-dimensional "Newtonian" container as with models of solar systems; or "curved" Einsteinian space as represented by computer projective representations.

Analysis of Variations

Although I have more suggested than developed the examples, I can now turn to a preliminary set of comparisons between three different cultural contexts with respect to the three variables introduced. The first group I shall call the *embodied set*. This set is illustrated in the archaic tropical astronomical and navigational systems. In this context bodily position and perspective are merged, either explicitly or implicitly. Now what this context grasps is a very fundamental and primitive experience: *I am my body.* And at this fundamental level, I may say that I experience the world from this "being-here" of bodily existence. The interpretation of space-time which is associated with this observational position and perspective can be and is coherent, orderly and while relative not "subjective." There is also a sense in which this cultural context is one in which the world is "Aristotelian" in that it is made up of concrete material individuals (sun, moon, stars, etc.) and direction is taken from these.

In the second cultural context, which I shall call the *disembodied set,* bodily position and perspective are separated. Indeed, the separation is such that a nonoccupied or hypothetical position is frequently given privileged status, usually in the form of some version of a bird's-eye perspective. Thus in both European navigation and in temperate area astronomy, an emphasis emerges upon what may be attained through a fixed, disembodied perspective. Associated with this context is what one might call a "platonic" world in that hypothetical dimension becomes the most important for the interpretation of space-time. The world-as-grid (the imaginary longitudinal and latitudinal lines on maps and globes) becomes the most important for locating the "accidental" location of our bodies. Space-time here anticipates what becomes the Newtonian universe which, historically, comes much later than its phenomenological position which was latent from the beginning.

But I have also suggested that contemporary astronomy (although not so much navigation) differs from both the previous sets in that at least implicitly it is multiperspectival. What is seen of the universe is relative to the perspective from which it is viewed. Thus only very recently have we come to understand that our very galaxy is four-armed and that we are located slightly above the equatorial mean of the Orion Arm.

While there remains, within what I am calling this Einsteinian context, a persistent use of the disembodied perspective, its very relativity recalls something of the embodied emphasis of the tropical context. The multiplicity of perspectives possible in contemporary astronomy—increasingly made materially possible by space technology and thus no longer merely imaginary—nevertheless also increasingly take account of the need for recognizing the position of the observation. Embodiment returns to the Einsteinian context precisely in the multiplicity of perspectives. Thus contrary to the possibility of taking bodily position as purely arbitrary, while not returning to the unity of position and perspective, the Einsteinian context is one which takes seriously whatever implied position and perspective is taken. We have here something like a phenomenological ascent in which the multiplicity of profiles begins to suggest something like a structure to the object of inquiry, but a structure arrived at by multiplying perspectives.

We have, then, three variations upon our variables of bodily position, perspective and space-time interpretation. But at this point it may seem that we are distant from culturally embedded technologies. I suggest, however, that this is not the case, for associated with each of these cultural sets are particular technologies (or lack of same) and particular uses which correspond to the points analyzed. One thing distinctive about Pacific navigation was in fact the absence of instruments. Waves, bird flights, stars, reflections in the sky, etc. are directly and perceptually read—a kind of body to earth-sea relation. The importance of this for the embodied context becomes apparent with the subsequent introduction of Western instrumentation.

The compass, recently introduced, is now a fairly common fixture upon sailing canoes, although it is used differently than by European navigators. We already noted that tropical space-time interpretation is not polar, and this remains the case with their use of the compass—that its needle points north/south is not a matter of importance for Puluwatians, only that it is a relative constant. Because it is easy to read; less affected by local conditions (such as storms which temporarily disrupt wave patterns), it "distances" and makes easier the task of steering. It lacks its role in a context of world-as-grid. But is very like the steel axe which is so easily accepted by stone age cultures unaware that it is not just an object, but a whole complex of intercultural relations. The compass makes for easy function and for the introduction of a new kind of distance. It transforms a perceptual context. But at this level, the instrument only inclines in a direction and does not determine it.

Tropical astronomy does utilize instruments—but curiously they remain "Aristotelian" in that they are specifically and purposely attuned to specific natural phenomena. South America abounds with "telescopes" which are permanently fixed for the locating of solstice or equinox. These are hollow tunnels built, for example, into tombs or monumental buildings, which allow the light of the sun at the right moment to shine on some particular designated spot. They are permanently fixed, single purpose instruments. Yet even here, they implicitly point to a geometrizing of the world.

As annual cycles are noted and accumulated, measurement of more and more phenomena lead to the sophisticated astronomical and calendrical systems, for example of the Maya. The material record and the material instrument display the possibility of a world instrumentally ordered. The instrument makes possible a different relation to space-time. Micronesian navigators do utilize imaginary or hypothetical entities in certain conditions. One such example is the use of "ghost islands" which, interestingly, may be considered to be points on the sidereal horizon from which to determine relative position (similarly, some parts of the sky are less populated with stars than others and this poses a problem for a strictly "Aristotelian" system). But the instrument, which may be divided equally into points, does not have this uneven but real distribution. Note that the instrument, then, transforms the *perceptual* situation in that *it* rather than the stars may be read. It distances from natural objects, but it becomes itself a perceived object to which and through which one may be located. Here, implicitly, is also a change of what counts as world. And, here, we achieve a first suggestion of variant cultural embeddedness of technologies.

The Einsteinian context is one which arose well after the rise of contemporary instrumentation. It entered history only after there was a technologically saturated culture. And I suspect that its multiplicity of perspectives with the associated accounting for the position from which the observation is taken, relates to another aspect of the human experience of technology. In the Einsteinian context the use of instrumentation is familiar, integral to scientific investigation itself, and experiences the technology as a kind of extended embodiment. The astronomer here sees *through* the instrumentation and, given the contemporary variety of instruments (telescope, spectrascope, space vehicles, satellites, etc.) the embodiments, too, are different. The multiplicity of perspectives is in this sense a familiar and taken for granted result of the mode of investigation.

At this stage, then, we have three different cultural contexts which also may be seen to be associated with three different technological complexes. So far, we may interpret each as a variant. But are there invariant features? And, further, are there detectable developmental features which may be derived from these variants?

Allow now a somewhat speculative turn with respect to technological embeddedness. If the embodied context is one which most easily associates with a direct perceptual reading of natural phenomena, what happens with the introduction of perceptual-mediating instruments? I have suggested that what first occurs is a kind of transforming distantiation. The instrument may both be read (as object) and read through (as extended sense). But this possibility is ambiguous and multistable. On its one side the instrument as object is other than me and that to which I must relate; on the other, it extends or amplifies my capacities such that it is partially *embodied* (Ihde 1979, 6–11).

Now my speculation is this: first or early uses of instruments must be learned. Before being embodied, they must be mastered and often that task is difficult. Take the automobile as an example: Most of us are so accustomed to driving that it is a kind of second nature—indeed, we pay little attention to the action, yet entailed in this embodiment of motion through the car is a good deal of familiarity with the car's capacities, traits, etc. Now feature the following situation, common to countries in which cars are used. A driver becomes a bit inebriated and, upon taking the wheel with both diminished senses and perhaps too extended sense of his ability to drive, soon is exceeding the speed limit. He comes to a curve, too fast, and the result is an accident easily explained in physicist's terms. Now imagine the two following reports: (a) "I guess I had too much to drink. In any case I got to the curve and found I was going too fast and lost control of the car. I ended up in the ditch." Or (b), "I had been to a party and had a few drinks. When I got in the car I headed it home, but when the curve in the road came up, the car, it had its own idea, and I could do nothing with it. It drove itself off the road and wrecked me." In both cases the loss of control describes correctly what happens when the inertial force exceeds the friction and center of gravity for the car to be under control. But in one case the driver sees the situation as one of losing control; in the other the driver sees the car as taking on its own character. One description begins from an embodied experience of the car; the other from a more disembodied or otherness experience of the car. Now these two descriptions were, in fact, made by persons from two different cultures, with the second coming from an indigenous person for whom cars are clearly relatively new to use.

What I am suggesting is this: Both between cultures and within a culture, early experiences of new technologies are likely to have heightened ambiguous results. A given technology may be both experienced as object tending toward the alien and/or experienced as tending toward a familiar embodiment. For example, in our culture the computer may be very differently experienced by those unfamiliar with it—I sometimes conduct seminars for parents with computer anxieties for themselves and their children—and by those familiar with it. Today's "hackers" are perhaps good examples of persons who have domesticated the computer under a new metaphor: *world as game.*

But even more, when new and especially complex technologies are introduced into variant cultures, the tendency is almost inevitably one of the alien side of the ambiguity. Here, finally, we return to cultural variants, and the Third World experience.

Third World and Technological Accommodation

Third world cultures, unlike the archaic examples used above, are contemporary cultures in which the expansion of Western science-technology is clearly directly experienced. I return here to the perception and interpretation of this impact. I noted that there is a tendency in the Third World context to see (a) science-technology as a unitary phenomenon, (b) as the expression of an alien culture, and to (c) experience a struggle with respect to some mode of accommodation which cannot be avoided.

In the first place, it may now appear that the perceptions of the unitary nature of the phenomenon are in fact true. Technologies are integrated into the type of knowledge or science utilized in the culture. Our science is embodied in a complex technology and is not even possible without it. But different complexes of technologies entail different interpretations of worlds. Second, the recognition of this fact speaks for the accuracy of the second perception as well. Western science-technology is a cultural expression, albeit a now highly powerful one. And, just as a steel axe entails a gestalt of dependency and interdependency relations, so today's high technology components entail similar gestalts. Only now our previously archaic cultures recognize this.

At a second level, one might speculatively note that perhaps some of the problem in technology transfer occurs developmentally. To move from areas of experience previously embedding no technologies, into those which now do entail technologies, calls for a massive cultural shift. This is dramatically the case with those technologies which entail intimate relations, for example, birth control. While there are some minimal techniques (not technologies) of child spacing—for example long lactation periods—many cultures still do not use and resist the use of technologies in this dimension of life. And here one can see dramatically how one change entails an entire gestalt: relations between women and men, kinship relations, values, hierarchies, and so forth.

At other levels, where technologies have been employed but where new technologies are introduced, there is a shift as well. The snowmobile and the rifle have dramatically changed Eskimo ways of life, yet the adaptation to shooting was a relatively easy one to make. There is also the problem of the essential ambiguity involved in learning the use of a technology. The multistability which allows a drift either toward a familiar embodiment, or towards a partially alienated otherness, is also sometimes culturally relative.

And at an even higher level, the movement and introduction of new technologies points to a kind of invariant which is itself cross-cultural. Here the example of computerization is a good one. The movement to a high-tech, computerized industrial complex is itself empirically virtually universal. And the problems which can be recognized in its wake, while differing in degree, are not totally dissimilar between the Third World and Euro-America. Probably the *telos* of computerization is such that it entails simultaneously low intense labor and high technical skills. The impact of such a movement upon countries where labor is both unskilled and lacking in intensity is obvious—but the same relation to the new style of unemployment in the United States can also be discerned.

It is, however, too early to make a telling prediction concerning the outcome of this movement, precisely because the technology itself is not yet part of the taken-for-granted way in which we may be embodied. We remain at the stage of fascination and lack the critical distance and familiarity to note just how dumb (or smart) this machine may be. We cannot even tell very well what uses are appropriate and what not. Although increasing numbers of people are beginning to learn about limitations and even debilities. Everyone experiences "the computer is down" as a delay to activity, and who can doubt that bank lines and transactions as well as those of ordinary sales transactions are today slower and more complicated than before? I shall never forget the prophetic memo passed around some years ago within the university which stated that we now must order our next semester's books at least four weeks earlier than previously because the ordering system was now computerized.

Contrarily, calculators have made our bookwork easier and my colleagues seem to love their word processors (as their tomes grow more Germanic by the day). Someday, hopefully, we will begin to learn appropriate uses for this machinery, although what uses these may be cannot be predicted.

Conclusion

What, then, is the moral of the story? Four things seem obvious. First, different cultures develop different technologies and uses of technology with respect to wider aspects of life forms. Negatively, technologies *per se* do not determine a form of life as such. Cultures with minimalist technologies and yet maximal centralist tendencies are possible, and vice-versa (at least in theory). In this respect technologies correlate with cultural outlooks.

But, neither are technologies neutral. Steel axes and compasses, unknown to their new users, allow different sets of habits to develop even if the artifact is not used in the same way it was in the donor culture. Thus the practice of wave-reading in Micronesia has declined in recent times and with it an erosion of an important element in perceptual navigation. Technologies incline when embedded in cultures.

Third, at virtually every stage of the development, use and adaptation of technologies, there is also an essential ambiguity or multistability possible for each technology. Any single technology can be used in ways not "intended"—this is the case for Heidegger. Heidegger's hammer, which can be used as a hammer, a paperweight, a murder weapon, or an art object—as well as for the computer network which can be used to convey money, keep book, or as a secret code for hackers to penetrate. Moreover, there are both embodiment possibilities (the car extends my bodily capacities and mobility) and objectifying possibilities (the car is "other," with alien personality).

Fourth, beyond variability, there are also stages in the adaptation and use of technologies as they are learned and made familiar. Early stages seem to entail fascination and fear and never reveal uses or consequences. Moreover, the fascination-fear stage is one which seldom reveals the limitations of the technologies involved. (I suspect that is where we are with respect to the computer as suggested.)

Where, then, does this leave us with respect to the global questions raised at the introduction? The answer has two sides, the first of which is cautionary. Given the multiplicity of possibilities, it is clearly premature to conclude that contemporary technology can or is leading humankind to a single cultural form. To answer that question one would need to know just how strong inclinatory directions are within the multiplicity of technologies we have developed may be.

Positively, it does seem clear that the empirical spread of contemporary technology is such that whatever variation there will be, will be one which utilizes a multiplicity of technologies—in short, in the clash and confrontation of cultures I see little evidence of trajectories toward minimalist forms. This is not to say that alongside the dominant cultural direction there might not be minimalist islands, just as in the nineteenth century industrialization there arose multiple sectarian revivals such as the Amish, the Amana Community and other forms of variation from the then current technological development.

Finally, our destiny clearly is tied to current technological trajectories. And while I have argued that we do not yet have anything like a whole-earth assessment of these trajectories, insofar as we are aware of this destiny, our task remains one of a most urgent need to inquire more deeply into technology and its cultural embeddedness.

References

Aveni, A. F. 1981. "Tropical Archeoastronomy." *Science* 213, no. 4504 (July 10): 161–71.

Gladwyn, Thomas. 1970. *East Is a Big Bird*. Cambridge, MA: Harvard University Press.

Heidegger, Martin. 1977. "The Question concerning Technology." In *Basic Writings*, edited by David Krell. New York: Harper and Row.

Ihde, Don. 1979. *Technics and Praxis*. Dordrecht: Reidel.

Mumford, Lewis. 1971. *Technology and Human Development*. New York: Harcourt, Brace.

Rapp, Friedrich. 1981. *Analytical Philosophy of Technology*. Reidel: Dordrech.

Chapter 10

The Designer Fallacy

Introduction

Imagine the following photograph: in it are two adolescent Amish girls, clad in the traditional garments with bonnets, dark dresses long and laced, but wearing *roller blades* (in-line skates) while skating along a small town sidewalk in Pennsylvania. This photograph has been publicised by the *New York Times* and other media recently. For the larger public, the response is likely to be one of a sense of incongruity between the traditions of horse and buggy, non-use of electricity, "plain clothing" and yet—roller blades as the latest, highest tech skates, equally popular among the lycra-clad, brassiere-showing urban youth in Central Park.

What the photograph does not reveal, however, is the process behind the Amish acceptance of this technology. As a colleague once pointed out to me, the Amish—in spite of the very technologically conservative approach taken by this religious community—have probably one of the most sophisticated and effective forms of *technology assessment* available. Every new technology is considered with strict considerations about whether or not it will support or enhance the values of that community, or detract from or erode those values. Thus, not everything—indeed not much—is accepted from the glut of current innovation that virtually immediately pervades the larger American or other industrially "advanced" societies. The roller blade, it turns out, plays the same role for the Amish as the little red wagon, the scooter, or other forms of human-powered entertainment. It can fit, it would seem, into the "plain" lifestyle of the Amish community in spite of its apparent incongruity for the rest of us. Electricity, television and cinema, and most other hi-tech technologies, remain excluded. I would only add that the technological conservatism of the Amish is not because of an anti-technological attitude, it is because of its deeply held (and conservative) religious beliefs.

I do not know if this community decision was a form of political compromise allowing Amish youth something new or not. It is known that many youth leave Amish society, yet the Amish have also expanded into areas from Ohio to upstate New York, beyond their previous boundaries as well. The tensions among the Amish are not dissimilar to the break-up of Eastern European Socialist countries in face of the onslaught of Western technological entertainment and consumer technologies that became so desirable, but unattainable or unaffordable in those societies.

The point of this vignette, however, is not to contrast hyper-consumerism and technological saturation with a religious form of minimalist nostalgia and communitarian values. I am using it, rather, as a hyperbolic indicator of a problem philosophers face with respect to changing technologies and the evaluation thereof. The typical role many think the philosopher ought to follow is that of "ethician" or the reflector upon normative aspects of technologies within societies. How *ought* we to deal with technologies? What will their effects be?

I do this because, from both within my own trajectory in the philosophy of technology, and increasingly from the recognition of other philosophers and historians of technology, there emerges a practical *antinomy* with respect to precisely the predictive problems in technological development.

The Philosopher's Prognostic Antinomy

The antinomy can be stated simply: if philosophers are to take any normative role concerning new technologies, they will find from both within the structure of technologies as such, and compounded historically by unexpected uses and unintended consequences, that technologies virtually always exceed or veer away from "intended" design. How, then, can any normative or prognostic role be possible?

Philosophers, typically, are expected to play post-development normative roles (as "ethicians" in applied ethics, for example). This usual role I shall call the "Hemingway role." That is, as Ernest Hemingway reflected his experience in *For Whom the Bell Tolls*, his job in wartime was in the ambulance corps. He did get into the battlefield, he was actually wounded, but his task was to pick up the casualties. He was part of the battlefield "clean-up squad."

This metaphor is appropriate for the many applied ethics roles occupied today by many philosophers. These began at first in the context of the development of medical therapeutic technologies—for example, during the early days of kidney dialysis, at first scarce and expensive, philosophers, theologians and other non-medical personnel were called upon as a "civilian ambulance corps" to deal with the ethical problems. Much of this relates to "lifeboat ethics" of scarcity situations and concerned decisions about who should get the limited treatment. The reason why the "Hemingway role" fits is that the Spanish Civil War ambulance corps, together with the nursing staff, had to practise *triage* on the spot: (a) who was dead or could not survive? (b) who was possibly recoverable or likely to live? (c) and who was borderline and questionable for recovery or life? Depending on the severity of the battle, the borderlines for triage could shift upwards or downwards.

I do not wish to discount the importance of the "Hemingway Role"— or of applied ethics. These are clearly an improvement over the pre-modern form of clean-up process. After a medieval battle, and only after, did the clean-up squad arrive. Then, sometimes after stripping the dead and dying, the injured could be moved and cared for. Not only was the chance for recovery lower, but the wounded had to remain on the field, bleeding, until the battle was over.

But in both cases the metaphor points to the end-game role always played by the ambulance corps. The therapy and healing roles they played remained absent from the strategy rooms of the officers and military commanders, and further still from the political considerations which always lie behind, before, and in the occurrence of war itself. For applied ethics in this context, it is always after the technologies are in place that the ambulance corps arrives.

I have argued on numerous occasions that if the philosopher is to play a more important role it must not be only in or limited to the "Hemingway role." Rather, it should take place in the equivalent of the officers' strategy meeting, before the battle takes shape. I will call this the "R&D role."

A first response to this proposal might well be: but who wants any philosophers amongst the generals? The research and development team? The science policy boards? (The implication is, of course, that philosophers will simply "gum up the works." And the excuse will be (a) that philosophers are not technical experts, and (b) any normative considerations this early will certainly slow things down—a sort of "Amish effect." Of course, the objections, in turn, imply the continuance of a status quo amongst the technocrats as well, free to develop anything whatsoever and free from reflective considerations.)

It should therefore be noted initially that the antinomy I am pointing to arises primarily for "R&D role" placed philosophers. There is an advantage to be had from having to deal with already extant problems in the "Hemingway role" position.

But, first, permit me to sharpen the antinomy: in my own work I have argued that *all technologies display ambiguous, multistable possibilities.* Contrarily, in both structure and history, technologies simply can't be reduced to *designed functions.* I have claimed that there is a "designer fallacy," which functions similarly to the "intentional fallacy" in literature. That is, if the meaning of a literary work cannot be traced or limited to the author's intent, similarly, in technology, its use, function and effect cannot and often does not reduce to designed intent.

Heidegger's hammer is a simple example: a hammer is "designed" to do certain things—drive nails into the shoemaker's shoe, or into shingles on my shed, or to nail down a floor—but the design cannot prevent a hammer from (a) becoming an objet d'art, (b) a murder weapon, (c) a paperweight, etc. Heidegger's insight was to have seen that an instrument *is what it does, and this in a context of assignments.* But he did not elaborate upon the multistable uses *any* technology can fall into with associated shifts in the complexes of "assignments" as well. No technology is "one thing," nor is it incapable of belonging to multiple contexts.

The same obtains with complex technologies: email in my university was first used to transmit memoranda, then as a substitute for "phone tag," then even for chain letters (which the administration tries hard to prevent) and even the propagation of computer viruses. And, as Kittler has well shown, the typewriter (and one can add, the telephone) was originally designed as a prosthetic device to help persons with sight deficiencies (or the telephone as a sort of hearing aid)—uses that became at most marginal as the office soon transformed the secretariat through the typewriter and communications through the telephone (Kittler 1989, 105–207).

I argue that the very structure of technologies is multistable, with respect to uses, to cultural embeddedness, and to politics as well. Multistability is not the same as "neutrality." Within multistability there lie *trajectories*—not just any trajectory, but partially determined trajectories. Optics takes us into the micro- and macroscopic as the histories of telescopes and microscopes evidences, but optics remains within the boundaries of the light spectrum and did not, by itself, develop into the new astronomy, which now ranges from the gamma ray short waves into the radio wave long waves, thus revealing a wider world. Similarly, the external fulfilled intentionality of a Moon mountain scene carries with it not only the magnification of this external phenomenon, but it magnifies the motion of the observer holding the telescope and thus reflexively opens the way to a discovery of bodily micro-motion—a trajectory not developed by Galileo but implicit within his favoured instrument.

These complexities of multistability clearly make prognosis difficult, perhaps impossible if the aim is full prognosis. These are multiple intrinsic possibilities of the technologies. Historians of technology, however, tend to focus upon *effects*, and here there are two books to which I will refer, which make the case brilliantly for the unforeseen and unintended uses, consequences and side effects that *all* technologies produce.

The first book is *Why Things Bite Back: Technology and the Revenge of Unintended Consequences* (1996) by Edward Tenner. I do not have the time to outline in detail all the forms of a "revenge theory" of technological consequence—which include differences between *rearranging, repeating, recomplicating, regenerating, and recongesting effects* (Tenner 1996, 8–9). But the book is glutted with examples of each. His project began by reflecting upon a prediction made by "futurologist" Alvin Toffler, concerning the coming of electronic media. Toffler says, "making paper copies of anything is a primitive use of [electronic] machines and violates their very spirit" (Tenner 1996, ix). Obviously, we are all aware of the "paperless" electronic society we now inhabit! Tenner claims, "Networking had actually multiplied paper use. When branches of Staples and OfficeMax opened near Princeton, the first items in the customer's view . . . were five-thousand sheet cases of paper for photocopiers, laser printers, and fax machines" (1996, ix). I shall not even attempt to list the multiple examples of these revenge effects—two simple ones illustrate how a single technology, which necessarily belongs to a context of assignments, produces unintended and often revenge side effects:

- "At home cheaper security systems are flooding police with false alarms, half of them caused by user errors. In Philadelphia, only 3000 of 157,000 calls from automatic security systems over three years were real; by diverting the full-time equivalent of 58 police officers for useless calls, the systems may have promoted crime elsewhere" (Tenner 1996, 7). (In my own village on Long Island, the situation is bad enough that the Trustees are considering fines for each false alarm.)

- Another example cited by Tenner comes from the well-known phenomenon where temperatures in cities are always higher than the countryside, at first due to pavement, stone and concrete, which retain heat—but then "improved" by air conditioning, which shifts interior heat to the exterior. Air-conditioned subway cars spill heat upon the platforms, which are often 10–15 degrees hotter, so that a ten-minute wait for a ten-minute ride actually produces a heat gain to the rider!

- A final familiar example comes from the change of composition technologies—"repetitive strain syndrome" and carpal tunnel syndrome were rarely known in the days of the typewriter, but have escalated with the computer. The harder and slower strike of the former yielded to the faster, lighter "advance" of the latter and contributes to this contemporary ailment.

We now have enough examples to clarify the antinomy: technologies "contain" multiple possibilities for use, direction and trajectory—are essentially multistable—making clear prediction of effect, use and outcome difficult if not impossible. And, once in place, technologies produce in the context of the multiple assignments to which they belong unintended and often "revenge" effects, again difficult, if not impossible to predict.

The second book I wish to cite is *Naked to the Bone* (1997) by Bettyann Kevles, a history of medical imaging. Here, the unintended side effects arise from precisely what was discovered to be—only subsequently designed to be—a medical technology that finally revealed bodies as transparent. The history of the X-ray begins this process: Röntgen's discovery was publicised through his distribution of an X-ray photograph of his wife's hand-with-ring, showing bone structure under and inside the flesh. This technology, begun in 1896, was one of the fastest to acceptance of any process. But to obtain the images long exposures were first necessary—up to 70 minutes in some cases. Retrospectively, we now know the result: severe radiation damage.

This was not knowledge that occurred immediately—indeed, one of the early uses of the X-ray was deliberate exposure to treat acne and skin disorders under long exposure time! Yet, by 1911, documented cases of burns, cancers and even deaths had accumulated. By 1920, in the incident cited by reviewers and which I repeat here, "At a meeting of radiologists in 1920, the menu featured chicken—a major faux pas because almost every one at the table was missing at least one hand and could not cut the meat" (Kevles 1997, 48). The history of the very instruments that make "non-intervention" possible for exploring the body, but which cause side effects through the examination, continues to the present. This, too, is part of the unpredicted revenge effect.

This double-dimensioned prognostic problem is, I am arguing, more of a problem for persons playing roles in relation to prognosis—in our case, the "R&D role" philosopher.

Philosophers in the "R&D" Position

The antinomy clearly points to the difficulties of any normative, prognostic role. But before I make suggestions concerning how to lower these difficulties, let us take a look at a few historical examples of "R&D role" philosophical attempts. Interestingly, the examples I will cite do not primarily belong to normative activity, but rather to *epistemological* aspects of technological development.

They are, however, suggestive of a positive role for the "R&D" philosopher: The most sustained example of the "R&D role" is exemplified primarily in the case of Hubert Dreyfus (he is not alone, but I shall use him as exemplar): in the early days of AI (artificial intelligence), Dreyfus was called in as a *consultant* by the RAND corporation to analyse and critique the development of AI programs precisely because they were failing to deliver either as fast or as effectively as the proponents predicted. The result was an *epistemologically scathing* critique of the program, *Alchemy and Artificial Intelligence* (1965), followed by several editions of *What Computers Can't Do* (1972, 1993). At the core of the critique were *epistemological* considerations concerning how human bodies work in intelligent behaviour. While many took Dreyfus as enemy, later on second- and third-generation computer designers began to see the alternative model Dreyfus proposed as positive (among these, T. Winograd's "ontological design" programs in particular). And these results have now spread to a much wider front, evidenced in a very recent article in *Science* on "The Space Around Us," in which Italian neuroscientists have adopted "motor intentionalities" from Husserl and Merleau-Ponty into cognitive science (*Science*, July 11, 1997, 190). This example not only is an exception to the applied ethics role usually expected, but is also an example of a philosophical insight being incorporated into both science and technology developments.

The second example comes from observations I have been able to make while "Euro-commuting" the last few years to northern European technical universities. I have come to know a number of philosophers located in these polytechs—they are often lonely in the sense that often there are no philosophy departments as such, although in some cases there are "applied philosophy departments." These philosophers, however, often find themselves on interdisciplinary research teams and play precisely "R&D" roles. My visiting role is frequently a secondary one: I am asked to review research proposals and give advice and criticism. Examples of such programs have included "Herman the Bull," a genetically engineered bull who has human genetic components designed to lower lactose allergies for humans who drink milk, to the "hermeneutics of crisis" in medical instrumental displays wherein reading multiple instruments itself may determine a crisis. Here are philosophers (indirectly myself as consultant) engaged in situ at developmental stages. I applaud both these directions with respect to the "R&D role" I am advocating. But my examples are not primarily normative, and the prognostic aims are minimal.

Even these examples hide failures of prediction: Dreyfus, in effect, predicted that "Big Blue" could not have been developed, and contrary to my own expectations about "what can be done, will be done," "Herman the Bull" has been put to pasture "without issue," as the legal profession might put it. But we have now seen how philosophers have entered "R&D" positions.

Prognostic Pragmatics

The antinomy remains: both structurally and historically, technologies present us with multistable ambiguities that exceed the bounds of rational and even prudential prognosis. Yet, to leave the situation simply there is not only to invite a laissez faire technological politics, but also to rule out even the possibility of critical reflection.

I shall here, instead, begin to outline a set of prognostic pragmatics that could serve, minimally, heuristic purposes:

- If technologies embody, both structurally and historically, the possibilities of multiple uses and unintended side effects, and all instantiate these in particular fashions, then one exclusionary rule for prognosis can be advised: *avoid ideological (utopian and dystopian) conclusions.* A utopian version of this, cited by Tenner, is John von Neumann's 1955 "prediction of energy too cheap to meter by 1980" (Tenner 1996, xi). A far less grandiose version occurred when philosopher of science Isaac Levi assured me that while he admitted that X-rays turned out to have harmful side effects, sonograms were bound to be totally harmless. Not more than a few months after this, I sent him a clipping about a study in Japan which indicated that frequently repeated sonograms seem to affect the central nervous systems of fetuses. Similarly, dystopian predictions include the worries of the nineteenth century over health effects of train travel—presumably so fast that it would cause heart problems. The "prediction" of side effects is not in itself dystopian, but pragmatically based upon long histories of similar side effects from all and any bodily intrusion—including non-radioactive ones. This is a generalised caution based both upon knowledge of the ambiguous structure of technologies and upon the related histories of similar instrumentation. In Kevles' history of medical imaging, it becomes clear that awareness of side effects has been amplified and that they are expected by today's practitioners. *No technologies are neutral, and all may be expected to have some negative (as well as positive) side effects.*

- From within the expectation that there will be side effects, a pragmatic caution might be: *if any negative effects begin to appear, amplify these and investigate immediately, err on the side of early caution.* In the X-ray case, skin burns were recognised very early, but techniques in shortening exposure time were slow in coming. It was also known that lead shielding prevented X-ray penetration, but shields for technicians were slow in coming. Similarly, King James (of "Bible fame") had already noted the noxious and negative health effects of tobacco in the 1600s—and we still do not have a safety standard for same.

- Technologies, unlike searches for theories of everything in science, thrive on alternative developments. *Enhance alternatives through multiple trajectories.* Here, energy production is a good negative example: R&D going into non-nuclear and non-fossil fuels has been scanty. In spite of this, solar development has become much more sophisticated and is finding wider uses—were R&D dollars deliberately directed towards a multi-source base, we might find more promising outcomes.

(In a forthcoming book, I demonstrate how contemporary sciences have increased breakthrough discoveries by the deliberate development and use of multivariant instruments. This "postmodern" multiperspectivalism in instrumentation has implications for technologies as well.)

• Design use experiments with non-expert and different users. The unexpected uses—both negatively and positively—of the Internet are interesting in this context. Negatively, our son's soccer coach from Vermont this summer was caught in a net sting as a paedophile; positively, we found a rather idyllic isolated ranch run on solar and cellular power through a travel page on the net. The net, interestingly, has displayed a respect in which dealing with technological prognosis is very like dealing with pornographic issues—that is, issues of freedom of expression, but related to idiosyncratic attractants, makes it extremely difficult to evaluate.

These heuristic suggestions are clearly not meant to be exhaustive. They are, at best, suggestive. Moreover, they more guide one in terms of what parameters to expect, but cannot determine particulars—but this problem is no worse, or better, than any other form of prognostic activity. They do imply that (a) we need to have a deep insight into both technological structure and the history of technologies—best based on broad and interdisciplinary knowledge; (b) that a critical take is called for, neither detracted by utopian nor dystopian aims; and (c) that multiple variant approaches are likely to be the most promising for contemporary complexities. And, it is my suggestion that philosophers seek precisely those situations that allow the expansion of the "R&D role."

References

Dreyfus, Hubert. 1965. *Alchemy and Artificial Intelligence*. Santa Monica, CA: Rand.
———. 1972. *What Computers Can't Do: A Critique of Artificial Reason*. New York: Harper and Row.
———. 1993. *What Computers Still Can't Do: A Critique of Artificial Reason*. Rev. ed. Cambridge, MA: MIT Press.
Kevles, Bettyann. 1997. *Naked to the Bone: Medical Imaging in the Twentieth Century*. New Brunswick, NJ: Rutgers University Press.

Kittler, Friedrich. 1989. "The Mechanized Philosopher." In *Looking After Nietzsche*, edited by Laurence A. Rickels. Albany: State University of New York Press.

Tenner, Edward. 1996. *Why Things Bite Back: Technology and the Revenge of Unintended Consequences*. New York: Knopf.

Part 3

The Phenomenology of Science

Chapter 11

The Critique of Husserl

Preface

This paper is a look at Husserl's explicit philosophy of science in the light of contemporary analyses of science in practice. *The Crisis*, published in 1936, was his last, major publication on this topic. Yet, it takes very little imagination to realize that since 1936, epochal changes have occurred in both the sciences and the interpretations of science, including philosophy of science.

Twentieth-century history of science, I would argue, has been marked by changes which are at least as profound as those which marked the turn from Aristotelean science to early modern science in the sixteenth and seventeenth centuries. Since WWII, the sciences have increasingly become Big Sciences, the term for which was coined by Derek de Solla Price in a book of that name (1963). Late modern science, first with chemistry, then physics and engineering, but today the biological sciences as well has become a science of corporate groups; of large state funding; of complex technologies in instrumentation; and implied major social-political dimensions for its operations. Husserl did live long enough to see and to some extent appreciate that in his recognition science had become "research."

Conceptually, there is as much distance between Newton and Einstein-Bohr, as earlier between Ptolemy and Copernicus. The interpreters of science, at first primarily historians and philosophers, began to catch up to these changes in the 1950s and 1960s, well after Husserl's death. The move away from science as theory machine, and the earlier notions of unified science, verification and universal law occurred with the "new philosophies of science" proposed by Kuhn, Feyerabend, Popper and Lakatos. And, they too, to some extent were propagating a notion of science as a research process which engaged corporate structures, vast technologies of instrumentation, megafunding, and the major social-political role which characterizes Big Science, as well as recognizing science as no longer the unified, generalized process early twentieth-century philosophers took it to be.

Nor did the "new" philosophies of science have the last word. The emergence of a virile set of new social sciences perspectives upon science, perspectives which focused upon science praxis, laboratory life, and the instruments or technologies of science also began to flourish later in the twentieth century.

Pre-WWII philosophy of science was simply superceded by the late modern turns of post-WWII science and its now multiple interpretations. Thus, we can expect that Husserl's philosophy of science will turn out to be, in significant respects, part of "history."

Where Was Husserl?

By this question, I mean to locate both in historical time and from Husserl's chosen perspectives, the outlines of how Husserl interpreted science and philosophy of science. Biographers and historians interested in Husserl know that from his own background and discipline, logic and mathematics were the primary cognate disciplines mastered by Husserl. And, so far as the sciences go, geometry and physics, with very small asides to astronomy, gain the most concentrated attention. The same may be said of his social engagements with scientists and philosophers of science of his day. The Göttingen group is an obvious standout, as were the mathematicians around Hilbert.

What kind of sciences are these? First of all, they are the most abstract of sciences, the sciences most given to formalisms, to mathematization, to idealization. Second, they are the least technologized of the sciences [excepting physics, of which more later] at least in the early twentieth century. Third, they are the sciences which least directly engage the whole body, or embodiment practices.

And, fourth and finally, they are the sciences which appear to be the most ahistorical insofar as they are expressed in the special language of mathematics which must be learned in special ways by all cultures approaching mathematics.

Yet, these were the paradigmatic sciences in the background for Husserl and for his symbolic spokesman for early modern science, Galileo Galilei (one could add that not only is this what Husserl sees when he talks about the sciences, but that this is also what he himself knows best in terms of science praxis).

Before turning to what is revolutionary and radical in Husserl's philosophy of science, I shall first take a few perspective glimpses of the state of the art during his productive apogee. By first turning to an abbreviated history of twentieth-century philosophies of science, one can better locate Husserl's site.

Twentieth Century Philosophy of Science in Three Steps

Early twentieth-century philosophies of science focused clearly upon what philosophers saw as the mathematization process which they found paradigmatically in the disciplines of physics and, in a more concrete sense, in astronomy, both of which were in effect a refinement of early modern epistemologies of the so-called "geometrical method."

Mathematizers

At the time of Husserl's early thoughts on Nature, for example in *Ideen II*, the three most prominent philosophers of science were Jules Poincare (1854–1912), Pierre Duhem (1861–1912), and, particularly for Germany, Ernst Mach (1838–1916). Amongst these, there was a nuanced consensus about the nature of knowledge and the role of mathematization. First, (a) all held to the basics of early modern epistemology, which has its subject-in-a-box (body) viewing its sensations which can only infer to an external world through representations. (b) This in turn leads to the notion that science is physical theory and its scientific exemplar is physics. Reasons for this narrowing of the sciences include the fact that most of those doing philosophy of science were themselves physicists or philosopher-physicists. Additionally, given the theory-bias of early twentieth century physics, physics as a discipline was more inclined to self-referencing. (c) Third, theory takes its shape through mathematization, whether this process is considered as a formalism (Duhem) or as instrumental (Mach and Poincare). Thus, science is the process of mathematizing the world through theory, and is paradigmatically exemplified in mathematical physics. Husserl, very much part of this milieu, thus also characterized science in its essentially physics mode, in a formalistic and abstract way. This was later to become part of what in the *Crisis* was characterized as science's forgetfulness of the lifeworld.

I think it is rather easy to see that it is precisely this "standard view" which Husserl himself takes for granted in his own characterization of at least the way science takes itself (in the early twentieth century). It was a rather "metaphysical" view of science, highly generalized and virtually non-empirical. Part of my thesis here is that Husserl is simultaneously both highly conservative and yet also radical with respect to his view of science. This taking for granted of the then standard view of mathematized physics as standing for science overall (*uberhaupt*) is Husserl's conservative moment.

Positivists

The second movement in the philosophy of science, was Logical Positivism or Empiricism, centered in the Vienna Circle in the 1920s. Within this movement, the consensus included (a) the notion of a unified science, that is, all sciences, ideally, would ultimately be unified in a single model, but such a unified science would also be reductionistic, ideally all other sciences would refer back to or be subsumed under such a physics. (b) A unified science would also be that which could progress through a hypothetical-deductive process, whose logic and propositional structure would be the task of philosophers of science to clarify and outline, with, once again (c) mathematized physics as favored model for operational sciences. [Those of you who are historically aware may recall that in the process of "purifying" science in this way, some members of the Circle in effect decided that such fields as geology could not be a genuine science since it did not well follow the hypothetical-deductive model!] The one addition which Logical Positivism-Empiricism did make to early twentieth century philosophy of science which would place it necessarily closer to Husserl, was the addition of (d) a verification process whereby observations must play a much more important role in science per se. Observations implied a more important perceptual role than had been taken in the earlier mathematized "pure" science.

Husserl himself saw the parallelism with Positivism in proclaiming that "we [phenomenologists] are the genuine positivists."[1] His view of science remained clearly that of a unified science, science *uberhaupt*, and not that of some set of discrete and often radically distinct praxes called sciences.

And, at the core of science remained the trajectory—which in his own way he questioned—which was dominantly theory driven, driven by the ascendancy of abstraction, formalism, and idealization. Husserl's philosophy of science, however, was also a radical variant compared to Positivism. The notions of lifeworld and of the bodily-perceptual plena were not the same as verified sense data.

Anti-positivism

The third moment in my quick three steps through philosophy of science occurs in the mid-twentieth century and begins what Peter Galison now calls the "positivist-anti-positivist" controversies (Galison 1997, 787–97). The anti-positivists include, pre-eminently, Thomas Kuhn, but also Paul Feyerabend, Imre Lakatos and Karl Popper. Oversimplifying, one can see in retrospect that each in his way (a) introduced much stronger strands of discontinuity into the interpretations of science, whether these were historical revolutions or paradigm displacements (Kuhn), or discrete research programs (Popper and Lakatos), or a series of ad hoc procedures not marked by unity (Feyerabend). (b) These anti-positivists also began to introduce more history as such into the image of science, and to a lesser extent, particularly with Kuhn, some sense of the role of instrumentation. (c) But, one could continue to claim that the anti-positivists retained the dominance of theory-production as the core of science activity.

Husserl, in relation to this third moment, was by mid-century, no longer alive. Thus, any relation to Husserl has to be by way of presumed anticipations from his work to theirs.

Husserl in All This

If one can speak of Husserlian hermeneutics, one would always find it circuitous. Husserl begins with some set of taken-for-granted beliefs, in this case the mathematization view of science in which the turn to a fully quantifiable analysis of nature constitutes the essence of science itself. He then circles backwards, downwards, and asks what science itself takes for granted or must hold implicitly. In the *Crisis*, this is the unique and peculiar form of historical derivation from a lifeworld to any particular science. Husserl's process is a "deconstruction" followed by a "reconstruction." But, in order both to foreshorten this process and to invert it, let us begin with the conclusions which Husserl claims to have reached by his questioning back to "origins."

First, what is basic, necessarily presupposed, and titled as "prescientific," is a lifeworld which is (a) bodily-perceptual,[2] (b) filled with practices,[3] and (c) cultural-historical.[4] All questioning back simply arrives at this fundament.[5] Thus, the most general claim which can be made about science is that it arises from, is dependent upon, and presupposes a perceptual, praxical, historical lifeworld.[6] Put into contemporary language this is equivalent to saying that science is not ahistorical, non-contextual, acultural, but rather is thoroughly historical, contextual and cultural. As we shall see later, this places Husserl squarely in the midst of the current discussion swirling within the "science wars."

Second, how is science different from the lifeworld? If it is different, does that imply a distance from the lifeworld? Here is a nexus of so much of the problem which arises for phenomenology in relation to the philosophy of science. The standard misinterpretation must be laid right at Husserl's door. It comes from his version of the symbolic first modern physicist, Galileo. Galileo, in inventing modern science forgets the lifeworld and its perceptual, praxical, historical origins, and distances and substitutes an idealized, abstracted world of mathematics for the fundament. The distancing of science from the lifeworld is a "forgetting" of its foundations and a "substituting" for it a "scientific world." The complexity of the lifeworld contrasts with the ideality and abstractness of the scientific world. It is here that science is described as in increasing process of abstraction, formalization, and idealization which leaves behind the plenary, perceptual, material bodies of the lifeworld.

Husserl's counter to this movement is his "invention" of a praxical world of geometry. In the origins of geometry, scientific idealization is the process of forgetting that the abstractions are abstractions from concrete practices of measurement—with instruments, etc. That is what presumably lies in the perceptual, praxical, historical "pre-scientific" lifeworld of geometry.[7] There is an upward, slippery incline of approximations into an ideal world which distances the investigator from the bodily-materiality of the lifeworld.

To be fair to Husserl, however, one has to note two qualifications which maintain a possible re-connection to the lifeworld. The first is that the very process of idealization and distancing from a primal lifeworld, is not simply negative, but is the attaining of a new level and region of autonomy which is positive so long as its origins are not forgotten.[8] And, secondly, because the lifeworld cannot simply be replaced, the process of idealization is characterized by Husserl as an indirect mode of analysis. It is an "application" rather than the uncovering of a platonic intrinsicality within the object analyzed.

In terms of the history of the philosophy of science, this Husserlian hermeneutic has been both the source of much pain for subsequent phenomenologists in this field, and yet also the location of Husserl's radical, as compared to his conservative dimension. One common misunderstanding which has doomed post-Husserlian phenomenology to unbased attacks concerning phenomeonology's subjectivism, is that phenomenology only deals with a pre-scientific, sensory or perceptual world, whereas science deals with the scientific or objectively constituted world. Rather, Husserl, as Merleau-Ponty after him, was after the descriptive analysis of the very distance and constitution of the lifeworld in relation to the world of science and thus must deal with both the pre- and post-scientific worlds. Yet, having proclaimed the pre-scientific lifeworld as fundamental, it also set this mode of analysis on a trajectory radically different from the early standard views of science. Husserlian philosophy of science, well before the science wars, was dubbed "anti-scientific," in part because it does not presuppose the automatic priority of a "scientific" world. But its radicality, too, depends upon the role which praxis has in the very building up of a science.

The problem lies in the various distinctions and relations between the lifeworld and the worlds of science. From Husserl's point of view these include: (a) The relation of foundation or fundament to derivative or result. There can be a lifeworld without science, but there can be no science without a lifeworld.[9] (b) Then there is the relation of the general "whole" of the lifeworld—it is presupposed by any human praxis whatever—and the partial or specially autonomous practices of science. Its abstractness implies that it remains partial.[10] (c) But, the special and autonomous situation of such practices are nevertheless both differentiated from the lifeworld and in a particular sense, "transcendent" to it. And it is in relation to this last relation and distinction that Husserl places the role of mathematization.

Husserl Got It Wrong

This characterization of Husserl's understanding of lifeworld/science relations and distinctions in some ways sounds distinctly contemporary. To this point I have characterized only three moments in twentieth century philosophy of science. But by mid-century interpretations of science began to emanate from what was to become known as science studies. Here the canon includes figures such as Steve Shapin and Simon Schaffer, David Bloor and Trevor Pinch, today Donna Haraway and Bruno Latour. These contemporary historicizers, sociologizers and anthropologizers of science, i.e., the science studies folk, would certainly agree with Husserl that science presupposes and remains cultural, historical, anthropological all the way down. They would also sharpen Husserl's insights about the particularity of science practices by no longer leaving these at the level of a "metaphysical" abstraction as Science with a capital. There are, instead, simply a series of differentiatable sciences in the plural, related at best by "family resemblances." And, contrary to the reductionist philosophers of science who portrayed science as a logical-mathematical theory machine, contemporary science studies sees the sciences as multidimensional practices which include an internal culture, embodiment in machinery, and expressiveness in special "tribal languages," only part of which is mathematical. With not too much stress, a lot of Husserl could be made to fit better with this image of the sciences than he could with his own philosophy of science peers.

But—as I have indicated in my subheading—in a very serious way he also got it wrong, or, he got science itself wrong and that through his own reductionistic version of Galileo. With this claim we reach the turning point and give Galileo back his instrument—the telescope.

Galileo with His Telescope

What follows will not deny the importance Galileo attached to the "mathematical language of nature" which he claimed.[11] Nor will it deny that in his own self-interpretation, Galileo fit into the distrust of the senses which were rhetorically popular in his time. But it is to ask a set of questions about what differences—and in particular what differences are made with respect to the lifeworld—occur if Galileo has a telescope.

First, let us turn historical in a stronger sense than the history practiced by Husserl, historical in a more historiological sense:

- We know that at late as 1597, Galileo is defending a Ptolemaic version of cosmology, a geocentric system of circular orbits.

- But by 1609 he has heard of Lippershey's compound lens telescope from a Jesuit friend, and given Galileo's own lens grinding skills, he re-invents his own version of a compound lens telescope which immediately improves upon Lippershey's mere 3× to Galileo's 9× version (by the time Galileo quit making telescopes, some 100 of them, he had improved to 30× which turns out to be the limits of lenses without chromatic distortions [Boorstin 1985]).

- And, although he took quite some time before he turned the new device to the heavens, once he did, he began to flood the world with his discoveries. The four brand new "firsts" which are usually credited to him include: (1) mountains and craters on the Moon, the sizes of which he estimated and which turned out to be taller than the Alps with which he was familiar; (2) the phases of Venus and (3) the satellites of Jupiter, both of which were crucial as confirmations of Copernican as opposed to Ptolemaic cosmology, which he henceforth defended; and (4) sunspots about which he wrote and which gained him his first warning from the Inquisition.

- The standard textbook history credits telescopic observations with the first observational confirmations of Copernican theory and Galileo as the holder of this honor. But, more, Galileo—never one to be humble—immediately proclaimed these new knowledges by creating the first science magazine, *Sidereus Nuncius*, usually translated as *The Heavenly Messenger*, which claimed sights not seen by Aristotle or the Church Fathers—and by extension, the Bible. This, too, would get him into trouble. But it was, of course, the mediated and magnified vision which radically differentiated his vision from that of Aristotle.[12]

- And the textbook history goes on to credit Galileo's observations with the destruction of the Aristotelean terrestrial/celestial physics distinctions by making Galilean physics "universal" or operative on both terrestrial and celestial planes.

Now, four centuries later, Galileo's telescope turns out to be both more radical than it was thought to be in his own time, but also less radical in terms of discoveries. The Jesuits, not without reason, argued that since so many telescopes provided double and even triple "images" that much of what Galileo claimed was simply "built into the instrument" as what we today call an "instrumental artifact" and thus could not be trusted.[13] Galileo himself was aware of this problem, and by claiming that any man could see what he could see, and in the process see more than any ancients, also noted that only after the observer was taught by Galileo could this result be assured. And, as firsts, Eurocentrists will probably be disappointed to learn that the scandalous sunspots were actually sighted and described as early as the ninth century by the Chinese. (How? One cannot directly observe sun spots since both magnification and light intensity blocking must occur. But the Chinese first lenses were made from dark quartz and that is how they saw the sunspots—their first eyeglasses thus were also the first "shades.")

If science is that human practice which yields new knowledge and knowledge which exceeds the bare or unaided senses, then only the science which is embodied in instruments which amplify and magnify ordinary capacities will qualify and that is the science produced by Galileo-with-his-telescope. And from even this scanty history, one should be able to see that much of the Galilean invention of early modern science took place only because Galileo did have a telescope, an artifact which receives no Husserlian mention.

Husserl's Ahistorical Galileo

If we now return to Husserl's Galileo, and better, to Husserl's claimed history of his forgetfulness of the lifeworld, we can see that Husserl really did not need a history at all to make the claims he made about Galileo's forgetfulness. What Husserl tries to reconstruct—presumably historically—is what Galileo had to take for granted concerning mathematics in order to invent the new science he was claiming. And what was taken for granted was the set of attainments which, originally based upon praxis and perception, become idealized and taken for granted in mathematical procedures which forget their praxical basis. What is sedimented, in other words, is the second order mathematical praxis which now leaves out of consideration the perceived, plenary bodies and objects we originally encounter, and which leaves out of account the attainments through measuring practices. The same points are made by Husserl in his "invented" and fanciful "histories" of geometry which originate in Egyptian survey practices or in Husserl's carpentry examples of creating ever smoother surfaces and shapes contrary to what actually happen in Galileo's history. Husserl's fanciful "histories" are imaginative variations which distance the multi-dimensioned plenary perceptual objects from the idealized and abstracted secondary objects of geometry and mathematization and calls the distance one of "forgetfulness." The "history" of this process is a history of meaning-acquisition. But Galileo's perceptions and practices with and through the telescope are precisely the measuring actions and perceptions which constitute the new "scientific" astronomical world.

But Husserl's "histories" are not histories in a second sense as well. If and when the Egyptians undertook surveying practices to establish field shapes, they did so with the praxical engagement with artifacts, technologies, just as when Galileo experimented, he engaged inclined planes, swaying pendula, and telescopes. But these material entities remain likewise absent from Husserl's "histories." Husserl's Galileo still needed a telescope, but now also an inclined plane and a swaying chandelier in the Pisa Cathedral.

Because Husserl Forgets Technologies, His Galileo "Forgets" the Lifeworld

One of the hermeneutic techniques I have developed over the years of reading our godfather's texts is that an author's illustrations not only embody the theories being developed, but they also often show the prejudices involved. For example, I challenge anyone to find a modern technology extolled, praised by, and extended in a positive way of revealing Being in the writings of Heidegger! Peasant's shoes, hammers, stone bridges, Greek temples, jugs, windmills, and hand held pens, all do serve positive and deep purposes. But steel bridges, nuclear plants, typewriters, hydroelectric dams, are all bad. In the face of this symptomatology, I find those defenders of Heidegger who claim he isn't actually anti-modern technology hard to believe. What kinds of examples do we find in Husserl? First, psychological, particularly perceptual-psychological examples abound to good use: listening to musical tones, tactile examples from the hand, memory examples of protension and retension, certain visual examples, abound to good if usually brief purpose. Then, there are the objects-before-one examples, and the already noted carpentry examples of smoothing and shaping. But—and this is a heuristic question—where are the instruments? The tools? The artifacts which are productive of change, insight or transformation?

His Galileo fits into this symptomotology: his Galileo is not the lens grinder, the user of telescopes, the fiddler with inclined planes, the dropper of weights from the Pisa Tower, but the observer who concentrates upon, on one side, the already idealized "objects" of geometry, and on the other the plenary ordinary objects which are before-the-eyes but indirectly analyzed into their geometrical components. Husserl's Galileo lacks the very mediating technologies which made his new world possible.

Writing

I am contending that Husserl's Galileo is a pre-selected and reduced Galileo, a Galileo without a telescope and were this Galileo to have had a telescope there could have been a radically different analysis of science and the lifeworld. And, at one point, Husserl himself came close to an insight which would have made a different analysis possible. It lies in the very brief remarks Husserl made about the "history" of writing.

The important function of written, documenting linguistic expression is that it makes communications possible without immediate or mediate persona address; it is, so to speak, communication become virtual. Written signs are, when considered from a purely corporeal point of view, straightforwardly, sensibly experienced; and it is always possible that they be intersubjectively experienced in common. (Husserl here contends that writing sediments meanings and as all sedimentation, the presentation is thus passive, but in reading the re-awakening of signification is an activation.) Accordingly, then, the writing-down affects a transformation of the original mode of being of the meaning-structure, within the geometrical sphere of self-evidence, of the geometrical structure which is put into words. It becomes sedimented, so to speak. But the reader can make it self-evident again, can reactivate the self-evidence (Husserl 1970, 360–61).

This is perhaps as close as Husserl comes to identifying a material technology and its praxis playing a role in which meaning-structures are not alienated from lifeworld praxis.

It would be stretching Husserl to claim that this is much of a recognition of technological artifactuality in a mediating role in any very detailed way, but at least there is a recognition that materiality can, through its very material transformation, make meaning-structures available to bodily humans. But, in Galileo's scientific praxis, this is exactly what the telescope does.

The Telescope as Science-Lifeworld Mediator

Let us begin with a rather close comparison between Husserl's version of writing and the Galilean telescope. First, what Husserl recognizes is that linguistic meaning-structure can be embodied and is embodied in actual speech. Sound is sensibly experienceable and its meanings can be intersubjectively understood. But writing transforms the bodily presentation to a visible mode, and to one which is "fixed" or repeatable. In this sense it is doubly "materialized." The telescope, on the other hand, does not transform the sensory modalities—one can alternatively see the Moon with eyeball or with mediating telescope. And, at least before drawing or photography which also belongs to science's trajectory, it does not "fix" its object the way writing does. But, the magnificational transformation which now makes what was previously invisible visible, nevertheless now makes repeatable and intersubjective the newly discovered Moon mountains. In these initial ways, the Galilean telescope is an analogue to linguistic writing, a technology which makes meaning-structures reactivatable, repeatable, and intersubjective.

And while this is probably all that we can legitimately draw from Husserl, the opening of a notion that instruments, technologies, can serve as mediators of meaning-structures can be extended to whole groups of technologies. Recording devices can do for speech what writing does for textual language in that the playback can be repeated and is intersubjectively available for reactivation at any time. Similarly, photography combined with telescopy can make Moon images also repeatable and intersubjectively available, etc. In short, there are many ways in which meaning-structures may be made available in the Husserlian sense, now extended to technologies.

But, there is also a way in which Galileo's telescope exceeds writing. Writing is a double transformation—it transforms verbal sensory presentation into a visual presentation, but it also calls for special "hermeneutics" if it is to be read. Whatever referentiality it has, must be learned in terms of a set of language skills and thus the referentiality to a thing itself is not perceptually transparent. The text may be presented passively, as Husserl says, but it also must be reactivated, which brings into play the reading skills of the perceiver. The telescope, on the other hand, while it transforms the "thing itself," the mountains of the Moon, does so by way of an isomorphic magnification and the instrumentally perceived object retains perceptual transparency. Galileo with a telescope is considerably more than a calculator or mathematician. Galileo with a telescope is also a perceiver and a practitioner within a now technologically mediated, enhanced world.

And it is here that we need to return to the Husserlian claims that Galileo's mathematization process produces a "forgetfulness" of the bodily-perceptual lifeworld. My claim is that in practice regardless of rhetoric or publication, Galileo never leaves the lifeworld. It is what he sees, albeit mediated by the telescope, the full plenary richness of the Jupiterian "stars"

or the spots of the sun, which changes how he understands the newly opening universe of meaning-structures not available to Aristotle, the Church
Fathers or the Biblical editors. It is just that the telescope precisely through
its transformation of perception, makes dimensions of the newly enhanced
lifeworld open to perceptual-bodily experience.

Why Forgetfulness?
And Whose Forgetfulness?

The science/lifeworld distance which Husserl claims originates in Galileo
is admittedly enhanced by some of Galileo's own rhetoric but not by his
praxis. On the other hand, Husserl is also "forgetful" insofar as he ignores
the transformational mediation of the telescope within Galileo's praxis.

Galileo explicitly made two seemingly contrary claims about the telescope: First, he claimed that anyone could see the Jupiterian "stars" or the
mountains on the Moon and thereby see more and better than the ancestors.
And, he explicitly claimed that telescopic mediated sight was better than
unaided eyeball sight. But, he also claimed that this everyman attainment
was only possible under instruction, i.e., Galileo's.[14] This seeming contrariety
dissolves under a quick and easy phenomenological analysis:

Galileo takes up his telescope, aiming it at the Moon. The quasi-
transparent instrument, now literally in "mediating position" between Galileo
and the heavenly object, produces a profound set of transformations. These
include the magnification of the Moon such that for the first time details
of mountains, seas, and craters immediately are visible. This magnification,
however, is not without cost, because the Moon thus made visible now ceases
to be placed in its normal, expansive location within the vault of the heavens.
In short, what we today call apparent distance is a phenomenological result
of telescope use. Were Husserl to have analyzed this apparent distance with
the aid of his own intentionality notions, he would have noted that the
Moon as noematic correlate in relation to the observer's bodily position or
noetic correlate has reduced the spatial distance. And, it is phenomenologically irrelevant if this is described as the "Moon is closer to me" or "I am
closer to the Moon." Nor does this exhaust the instrumental transformation
of experienced spatiality. Just as the Moon mountains are suddenly "closer"
within telescopic magnification, so also are my own bodily motions magnified
in the use of the telescope. (Users hand holding ancient telescopes have to

learn how to "fix" the Moon, and that is part of Galileo's instruction. A tripod helps, but that magnifies the apparent speed of Moon motion and one has to constantly adjust the telescope to the moving location of the Moon. In all of this, there are lessons and knowledges which were learned in early modernity, but all because of the mediational experiences of science/ lifeworld praxes involving the technologies of mediation.)

We can even, in retrospect, now see how Galileo's telescopic praxis was itself selective. Galileo was much more interested in what lies "out there" in the motions of heavenly bodies and astronomy, than he was with his own bodily self-knowledge which I have hinted was equally made present by telescopic praxis. It was only much later that the reflexive instrumental knowledges became more interesting, yet today in simulation devices, the gradually growing sophistication of virtual reality processes, it is largely this bodily self-knowledge which is being enhanced.

The Perceptuality of "Theory"

If Galileo's telescope mediated science and the lifeworld, thus leaving Galileo in a lifeworld to begin with, there is one final step which I should like to take. Both Martin Heidegger and Thomas Kuhn have noted and commented upon how differently Aristotle and Galileo "saw" motion.[15] For Aristotle, the swaying chandelier in the Pisa Cathedral, was a case of "restrained fall" but for Galileo it was a "pendulum." Both Heidegger and Kuhn recognize that in this there is a serious gestalt difference related to a complex background which is incommensurate between Aristotle and Galileo. And whereas Heidegger sees this as a disruption in the history of metaphysics and Kuhn sees it as a Wittgensteinian "duck-rabbit" gestalt shift, both are instances of how a thing is seen.

But this seeing is cultural and hermeneutic, not simply bodily-perceptual. Yet, this is the case as well for all seeing—without bodily and sensory sight there is no sight at all, but there is not mere seeing without its meaning-context within which it fits. Husserl's lifeworld does contain both of these dimensions, the bodily-sensory basis which falls under his descriptions of plenary things, and the history of praxis which falls under the notion of "origins." That he failed to take the synthesizing move which takes both as "perceptual" also plays a role in his forgetfulness of Galileo's telescope.

References

Boorstin, Daniel J. 1985. *The Discoverers*. New York: Vintage.

Brown, Harold I. 1985. "Galileo on the Telescope and the Eye." *Journal of the History of Ideas* 45:487–501.

Galison, Peter. 1997. *Image and Logic: A Material Culture of Microphysics*. Chicago: University of Chicago Press.

Heidegger, Martin. 1997. "Age of the World Picture." In *Science and the Quest for Reality*, edited by Alfred I. Tauber, 70–90. New York: New York University Press.

Husserl, Edumund. 1970. In *The Crisis of European Sciences and Transcendental Phenomenology*, edited by David Carr. Evanston, IL: Northwestern University Press.

———. 1982. *Ideas Pertaining to a Pure Phenomenological Philosophy: First Book: General Introduction to a Pure Phenomenology*. Translated by F. Kersten. Dordrecht: Springer.

Kuhn, Thomas S. 1962. *The Structure of Scientific Revolutions*. Chicago: University of Chicago Press.

Price, Derek de Solla. 1963. *Little Science, Big Science*. New York: Columbia University Press.

Sobel, Dava. 1999. *Galileo's Daughter*. New York: Walker.

Chapter 12

Technology Leads Science

Introduction

The thesis I wish to explore in this essay is that *there is a significant sense in which technology may be seen to be both ontologically and historically prior to science*. There is, of course, an obvious and trivial sense in which this claim may be regarded as true. If technologies in the broadest and most concrete sense involve humans and their uses of tools and artifacts, then at the least one can say that technology in this sense is both universal and probably used at the time of the arising of the human species. There are no instances of societies, cultures, or human groups which do not use tools and artifacts in their relations with the natural environment.

And if science centrally involves a theorizing about things in a systematic and hypothetical sense, then it should be apparent that the practiced and skilled uses of technologies long precede the kind of self-awareness implied in science. In the most general sense then, *praxis* precedes explicit theory.

I wish, however, to suggest that there is a more specific sense in which technology, particularly in its more recent developments, is the *condition of the possibility of science*. I argued in *Technics and Praxis* (1979) that science, in its contemporary sense as an experimental science wedded to specific meanings of measurement, is *necessarily embodied* in its instrumentation. Indeed, one of the chief differences between modern science and Greek contemplative science lies in the development of instrumentation both for measurement and for actual investigative purposes. I showed how instrumentation extends and embodies perception.

Historically, of course, even Greek science in actual practice engaged some measurement technologies. But the lack of a specific technological impetus also doomed Greek science to its primarily speculative attainments (witness the odd ideas about the shapes of atoms and causes of sweet, bitter, or sour tastes in Democritus. Lacking any means of investigation of such micro-phenomena, the speculation had to remain just that). This lack of appropriate technology determined the limits of a primarily contemplative science.

Here I wish to push the essential interlocking of science and technology further by arguing for the historical-ontological priority of technology as a condition of the possibility of science. I shall develop three unequal stages in this demonstration: First, I shall briefly describe what I take to be the standard and dominant theory of the relationship between technology and science. Second, I shall pay my debts to two important intellectual predecessors of my views. The philosophical debt is owed to Martin Heidegger, who may be said to have originated and solidified what has become the philosophy of technology for the twentieth century, and who argued most explicitly for the ontological priority of technology over science. The historical debt is owed to the large body of work done by Lynn White, Jr., who made us aware that there was a virtual technological revolution in the Middle Ages which preceded and laid the groundwork for the rise of modern science in the Renaissance and through the Enlightenment. The third step will then be an examination of certain aspects of the historical technological life-world. I shall develop this account along phenomenological lines.

The Standard Theory

There are a variety of conceptual possibilities which could account for the relationship of technology and science, but two extreme cases—I shall call them the "idealist" and "materialist" interpretations—have the advantage of posing the issues most starkly.

What I shall call the idealist view is the interpretation which holds that science precedes and founds technology. It is an interpretation which holds that requisite for creating a (modern) technology, one must have insight into the laws of nature, a conceptual system at the formal and abstract level, and the ability to *apply* this knowledge to the material realm, thus creating a technology.

In this interpretation, technology follows from science, both onto-logically as an application of scientific knowledge and historically as the spread of this insight into ever-widening realms of material construction.

The standard view is accompanied by an interpretation of the history of modern science and technology which may be characterized as follows: After a long dark period in European history, a revival of the Greek scientific spirit emerges within and animates what we call the Renaissance. Europeans regain an interest in nature, speculate about nature and evolve a method of understanding nature which we call Modern Science. Historically this movement becomes dramatic and fulfilled in such figures as Galileo, Kepler, Copernicus and eventually becomes fully systematized with Newton.

The rise of Modern Science is a development which includes (a) the discoveries of more sophisticated mathematics; (b) a gradual move away from religious and theological notions and a move towards a more mechanistic and materialist metaphysics; (c) a method which diverges from the more speculative ancient roots towards a more experimental and verification direction; and (d) a movement which results in the rise of physics as the primary science or at least the science which is first among equals.

Only after this historical development of science does there arise a technology (in the modern sense). The Industrial Revolution of the past century and a half and the explosion of the current "high technology" are plausibly dependent upon the precondition of scientific theory. Technology in the contemporary sense seems to spin forth almost directly from science itself.

In this essay I am not interested in a further exposition of the implicit metaphysics of this interpretation, nor am I going to undertake a direct attack upon its presuppositions. As an interpretation of the relationship between science and technology it has both plausible and implausible aspects. I shall point out some of these, but I shall do so indirectly by elaborating a strategy which this view must entail.

What must technology be, how must it appear if this view is correct? First, what will pass for technology must in the paradigm case be a technology which is obviously dependent for its shape upon scientific-theoretical considerations. Thus, the best examples are what we call today high technologies. While I do not intend what follows to be exhaustive by way of definition, I suspect a high technology must be characterized as a technology which must include: (a) a complex and interlocked system; (b) workings which are understood only by way of scientifically derived theories; (c) components which contain esoteric compounds and units, themselves the result of complex and scientifically determined processes; and (d) microscopic machine tolerances, internal organization, mechanical or electronic motions developed from micro-levels of manufacture and planned construction. A computer is an obvious case of such a high technology, but there are dozens of other examples which could do as well.

In contrast, "low" or, better, traditional technologies would be those which are simple, arrived at through a process of trial and error, which contain only rough interrelations of parts, and are understandable by any mechanically inclined person. A waterwheel is an example of such a technology.

That there is an apparent and even dramatic difference between the computer and the waterwheel seems clear. But just what and how that difference is to be accounted for is precisely what needs note. However, at the level at which I am developing the case, we need to be aware that the idealist position which holds that science is the condition of technology must accentuate a sharp difference between a presumed prescientific and scientific technology. In short, contemporary technology is seen to be disjunctive with traditional technology.

This tactic is conceptually necessary because otherwise one would have no way of accounting for the previously noted historical situation in which all peoples and societies use and have technologies whether or not they have a science in our sense. The historical dependence of technology upon science then becomes a special case of dependence; only *scientific* technology is historically dependent upon science.

The relationship of the Renaissance and Enlightenment periods to the Medieval Period may be seen to be an instance of the focus upon the assumed priority of science in the modern sense. Put most simply, because scientific knowledge as theoretical knowledge was assumed to be higher than so-called practical knowledge, the possibly unique attainment of the Middle Ages was overlooked.

A Materialist Theory: Heidegger and White

A contrary position is possible. I shall construct such a view by combining the insights of Martin Heidegger and Lynn White, Jr.

Martin Heidegger is perhaps the philosopher who has most originally and profoundly rendered the question of technology a central concern of philosophy. The position he developed in "The Question concerning Technology" is one which argues for the ontological, but not the historical priority of technology over science. The argument is complex and I shall look at only a few elements of it.

Heidegger holds that Technology has always underlain what we have called science in the West, but it has been revealed as the origin of science only recently. Embedded in this complex argument, however, is a deep ambiguity about what shall count as technology. On the ontological level, Technology—more precisely the essence of technology—is a certain way of experiencing, relating to and organizing the way humans relate to the natural world.[1] On the historical level, at least in the chronological sense, Heidegger seems to grant that technology in its modern sense is "later than" science. In short, Heidegger accepts in some degree the notion that modern or scientific technology is essentially and distinctly different from traditional technology. I hold that he is wrong in allowing himself to accept this notion, and as a result he weakens his own case in such a way to give credence to the usual accusation that he is somewhat "romantic" with respect to technology. In sum, the Heideggerian position is that Technology, while ontologically prior to science, is historically later.

At the core of the view which Heidegger is espousing, lies an inversion of the standard view of the relationship between science and technology. This inversion is most dramatically illustrated by his claim that rather than technology being a tool of modern physics, it is exactly the opposite: physics is the necessary tool of Technology. In this first instance, Heidegger discerns that modern physics is necessarily interrelated with its instruments:

> It is said that modern technology is something incomparably different from all earlier technologies because it is based on modern physics as an exact science. Meanwhile we have come to understand more clearly that the reverse holds true as well: modern physics, as experimental, is dependent upon technical apparatus and upon progress in the building of apparatus. The establishing of this mutual relationship between technology and physics is correct. But it remains a merely historiographical establishing of facts and says nothing about that in which this mutual relationship is grounded. (Heidegger 1977, 296)

Then, in a much stronger statement, Heidegger argues that physics is the herald of Technology: "Modern science's way of representing pursues and entraps nature as a calculable coherence of forces. Modern physics is not experimental physics because it applies apparatus to the questioning of nature. The reverse is true. Because physics, indeed already as pure theory, sets nature up to exhibit itself as a coherence of forces calculable in advance, it orders its experiments precisely for the purpose of asking whether and how nature reports itself when set up this way" (Heidegger 1977, 302).

This inversion, clearly evidenced in the way Heidegger views the relationship between science and technology, is one which nevertheless retains at least one partial sense in which science precedes technology. (I am quite aware, with most Heidegger scholars, of the distinction between *Historie* and *Geschichte* in Heidegger's use. However, *Geschichte* serves a specifically ontological function.)

This residual sense in which science historically precedes technology also accounts for a distinction between scientific and traditional technology. The strongest statement concerning this residual sense states: "Chronologically speaking, modern physical science begins in the seventeenth century. In contrast, machine-power technology develops only in the second half of the eighteenth century. But modern technology, which for chronological reckoning is the later, is, from the point of view of the essence holding sway within it, historically earlier" (Heidegger 1977, 304). Similarly, the disjunctive sense which the standard view must maintain and which separates modern from traditional technology, is allowed by Heidegger: "The revealing that rules modern technology is a challenging, which puts to nature the unreasonable demand that it supply energy which can be extracted and stored as such. But does this not hold true for the old windmill as well? No. Its sails do indeed turn in the wind; they are left entirely to the wind's blowing. But the windmill does not unlock energy from the air currents in order to store it" (Heidegger 1977, 296).

And again, Heidegger, as he so frequently does, contrasts the peasant's sense of earth from that of the modern technologist's: "In contrast, a tract of land is challenged in the hauling out of coal and ore. The earth now reveals itself as a coal mining district, the soil as a mineral deposit. The field that the peasant formerly cultivated and set in order appears different from how it did when to set in order still meant to take care of and maintain" (Heidegger 1977, 296). Thus while we have the assertion of the ontological priority of Technology over science as an inversion of the standard view, a secondary sense is retained in which technology chronologically follows the development of science and a sense in which there is a disjunctive difference between traditional technology and modern technology. Science, in Heidegger's view, stands as the event which finally shows to us what Technology is ontologically. Science is the herald of Technology in a (chronological) historical sense: "The modern physical theory of nature prepares the way not simply for technology, but for the essence of modern technology. For such a gathering-together, which challenges man to reveal by way of ordering already holds sway in physics. But in it that gathering does not yet come expressly to the fore. Modern physics is the herald of enframing, a herald whose origin is still unknown" (Heidegger 1977, 303).

What holds this argument together lies in the several ways in which Heidegger uses the term technology.

What may be called the surface definition of technology is what Heidegger calls the anthropological-instrumental understanding of technology, technology as a mere tool of science.[2] This definition, not false, is only merely correct. It does not reveal the *essence* of technology.

A second definition derives from the Greek *technē-*, and begins to more nearly approximate the Heideggerian sense of Technology in that *technē* is both a name for the activities and skills of a craftsman and for the arts of both mind and hand, but also is linked to creative making, *poiēsis* (Heidegger 1977, 294). For the Greeks *technē* was a production which was a kind of knowledge.

The third, and ultimate Heideggerian definition of Technology, however, makes of Technology a mode of truth or revealing (*alētheia*). Technology, in essence, reveals a world in a certain way. "Every bringing-forth is grounded in revealing" (Heidegger 1977, 294). "Technology is a mode of revealing, Technology comes to presence in the realm where revealing and unconcealment take place, where *alētheia,* truth, happens" (Heidegger 1977, 294). The essence of technology allows us to see, to order, to relate to the world in a particular way. Nature becomes standing-reserve, a source of energy for human use, and this mode of relating to the world becomes, in a technological era, the dominant and primary way in which we understand world.

I shall not further explore the Heideggerian view, except to note that only after Technology is discovered to be this way of relating to the world may one begin to understand how science, under this mode, is seen to be the necessary "tool" of Technology. Science becomes a means of knowledge which gives power, science becomes Baconian. And with this move the inversion is completed: Technology as the revelation of the world as standing-reserve is the ontological presupposition and ground of modern science.

Philosophically things would have been neater and clearer were it the case that Technology could be shown to be not only ontologically, but historically prior to science. And this would especially be so if the historical priority were of such a nature as to be understood as an experiential condition of the possibility of modern science. Such a view would also have the advantage that it would be continuous with the ordinary observation that some form of technology is universal and occurs wherever there are human societies.

I think this is the implicit import of the work of Lynn White, Jr., who has clearly caused a revision of the way in which we understand the Medieval Period with respect to technology.

White's publications concerning Medieval technology span two decades. The landmark book, *Medieval Technology and Social Change* (1962), shows how technological development was deeply implicated in systems of warfare (the stirrup led to mounted shock warfare, thence to changes in social structure in Feudalism), agriculture (one plough combined with horse power and the development of three-field rotation led to a shift of food production to Northern Europe) and in that increasing hunger for mechanical power which laid the basis for other forms of increased productivity.

By looking at the burgeoning technology of the Medieval Period, White paints a historical picture of a Europe rapidly changing, avidly searching for inventions; and particularly hungry for power. This is the case with the newly invented mechanical devices for extracting power from water and wind. By 983 water power was being used for fulling mills, but within a century the *Domesday* census revealed that there were already 5,624 water mills in operation in England (a harbinger of the Industrial Revolution centuries later) (White 1962, 84). The windmill was referred to as early as 1180 and was common in much of Europe by 1240. The search for power in the Middle Ages utilized every source. Inventions from foreign lands were rapidly experimented with in new ways, often in practical, but rarely overlooked. This medieval search for power laid the groundwork for later industrial technology but it was also intricately tied to a search for knowledge. Giovanni da Fontana, for example, in 1420, designed the forerunners of our robot measurers in the form of swimming fish, flying birds, and running rabbits, all linked to a plan to measure surfaces and distances in water, the air and out-of-the-way places (White 1962, 98).

One dramatic technological development during this period, a development which transformed the human perception of time, was the clock. In White's words, "Suddenly, towards the middle of the fourteenth century, . . . clocks seized the imagination of our ancestors. . . . No European community felt able to hold up its head unless in its midst the planets wheeled in cycles and epicycles, while angels trumpeted, cocks crew, and apostles, kings, and prophets marched and countermarched at the booming of the hours" (White 1962, 124). Time and the movement of the spheres was tied to a mechanical device. And thus by 1382 the universe itself began to be conceived of according to a mechanical metaphor: "It is the works of the great ecclesiastic and mathematician Nicholas Oresmus, who died in 1382 as Bishop of Lisicux, that we first find the metaphor of the universe as a vast mechanical clock created and set running by God so that 'all the wheels move as harmoniously as possible.' It was a notion with a future: eventually the metaphor became a metaphysics" (White 1962, 125).

White's more recent works have taken account of the unique intellectual climate which encouraged technological development in Europe. By the time of his publication, "Cultural Climates and Technological Advance in the Middle Ages," White can claim, "The technological creativity of medieval Europe is one of the resonant facts of history" (White 1972, 171). What he finds is that Medieval Europe was highly receptive to the use and development of technology and that several factors encouraged this: The organization and climate for order, stemming from the earlier Monastic reforms, readily adapted technology. The clock, used first to establish the order of time, agricultural techniques, and machines to lighten labor were all affirmatively valued. Indeed, his survey of the literature of the time finds that detractors from the praise of technology are rare. Contrarily, praise of invention, machines and their use is the rule.

Prior to our Bishop Oresmus who declares the heavens to be clockwork, one finds praise and prediction concerning a glorious technological future common: "Roger Bacon, 1260, pondering transportation, confidently prophesied an age of automobiles, submarines, and airplanes" (White 1972, 173).

This attitude of fascination and obsession with the technological stands in stark contrast to other areas of Christian civilization. Whereas the Latin West from the monasteries on accepted technology into the precincts of the holy—every cathedral must have a clock—the Eastern regions forbade such inventions in sacred space. Clocks must remain outside the realm of eternity, thus outside the church in the Orthodox lands.[3]

The positive evaluation of inventiveness, linked to a desire for machine power, was also accompanied by the willingness to adapt ideas and artifacts from any culture. What became the bow for our string instruments came from Southeast Asia. A Tibetan prayer wheel may have inspired the windmill, and so the list goes. In short, the Medieval Period was suffused with interest in, desire for, and the development of technologies.

By the late Middle Ages, at the dawn of the time for the rise of modern science, White points out:

> About 1450 European intellectuals began to become aware of technological progress not as a project (. . . this came in the late thirteenth century) but as an historic and happy fact, when Giovanno Tortelli, a humanist at the papal court, composed an essay listing, and rejoicing over, new inventions unknown to the ancients. . . . It was axiomatic that man was serving God by serving himself in the technological mastery of nature. Because medieval men believed this, they devoted themselves in great numbers and with enthusiasm to the process of invention. (White 1972, 199)

In short, White established that by 1500, a period whose image is consolidated by the technological genius of Da Vinci, there is a self-awareness of technology, the process of invention, and the desire to master Nature through human artifacts.

By the year 1500, Europe had already developed some of the instrumentation so fundamental to the very investigative possibility of science in the modern experimental sense. Lenses were invented by 1050, compound lenses by 1270, spectacles by 1285 and, by 1600, Gallileo's period, the microscope and telescope. Clocks, essential to measurement, began to be developed in the ninth and tenth centuries and by the 1500s were widespread from cathedral to town hall to individual watches.

On the industrial side one can note that Europe is by this time covered with wind and water mills; the low lands were being drained by wind power; there were railways in mines; and the massive, sophisticated architecture of cathedrals, suspension bridges and other large projects were part of daily life. Yet, in spite of the now reflective obviousness of this pervasive technological achievement of the Middle Ages, White is probably right in still claiming that, "the scholarly discovery of the significance of technological advance in medieval life is so recent that it has not yet been assimilated to our normal image of the period" (White 1972, 180).

The Historical-Ontological Priority of Technology

If one combines the claims of Heidegger concerning the ontological priority of technology with those of White concerning the immediately preceding historical technological revolution, one arrives at this essay's thesis. However, to consolidate this thesis I shall speculatively develop something of a phenomenology of daily life, first as it appeared in the European life-world, then as a variation, as it appeared in a different culture, that of the Polynesians. In so doing I shall focus upon spatial and temporal orientations.

A "Reconstruction" of an Aspect of the Medieval Lifeworld

My strategy in this reconstruction of a Medieval life-world will be to focus upon selected experiential components as they are embodied in praxis. It should be obvious by now that in the late Medieval Period mechanical contrivances were very common and indeed pervasive in many ordinary activities. The world was already implicitly thought of in terms of mechanical metaphors. But in my focus upon space and time I am concerned with the way these dimensions are *perceived*.

I begin with the familiar example of clocks, which were common in daily life in the late Medieval world. Lewis Mumford in his 1936 book, *Technics and Civilization*, has already noted how the clock was crucial to the development and reorganization of Medieval life. According to Mumford, clocks were first commonly used in conjunction with monastic life and the development of disciplined and common order. The keeping of hours for religious exercises and the ordering of work set the pace for public or intersubjective life. Heidegger, too, in *Being and Time* (1962; German original in 1927) pointed out the way in which clocks are not mere artifacts, but "take account" of human surroundings and nature. One can say that once clocks are developed, we begin to perceive time through technology.

Take careful note of the specific perceptual representation of time via the clock. First, until recently, all clocks represent time through a use of moving pointers. This is the case whether one regards the moving shadow of the sundial, the linear scale of the early water clocks, or the eventual round cyclical face of the cathedral clock. I would point out here that this representation of time is one which has both a focus—the instant of time which is the precise "now" as that point where the pointer "stands"—and a duration or span of time within which the instant finds its place. The field or span of time is the spread of the clock face, whether linear or circular. Thus "now" takes its place within a duration of time.

If one begins to reflect upon the evolution of the clock, one can note the following distinct developments: at first the movement of the pointer is crude and relates primarily to fairly large "units" of time. The earliest circular faces of clocks were marked only into hours and had only one hand. But as clockwork became more mechanically refined, time was divided into smaller and smaller units; a second pointer was added to mark the minutes, and then a third to mark the seconds. Time was more and more quantified. This quantification was gradually more finely divided and the perception of time became even more open to finer discriminations, to what may be called the micro-features of time. Moreover, these micro-features could be considered atomistically as units which were discrete from each other. In short, the clock allows us to perceive time latently as a series of atomized, discrete instants, a representation of what was to become a "scientific" mode of analyzing time. Time is perceived via or through the clock and this perception is a technologically mediated perception.

Historically, what eventually became more and more important was the focal point of technologically mediated time. The instant of its micro-features stands out. It becomes the means for further investigating things and is now essential for contemporary scientific measurement. Simultaneously, but almost unnoticeably, the field of time, which is the background but grounding feature of clock time, recedes and becomes less important. This development reaches a qualitatively different result in the contemporary invention of the digital clock. The digital clock represents only the focal instant of time, the field of time is no longer perceptually represented and in the process the perception of time also changes. The person who awaits the train, who once could glance at his watch and *see* that it was yet ten minutes until arrival time by *seeing* the relation between the pointers and the span, now sees only the number and must infer or calculate the span. This is to say that the mental operation for telling the time changes, even if unnoticeably, with the digital clock. What this portends for us, I shall not now predict other than to observe that if part of the essence of technology is "calculative thought" in Heidegger's sense, then the digital clock is an enhancement of this process.

Clocks were, prior to the rise of science proper, part of the daily experience of Medieval humanity. They were an ordinary part of the lifeworld, the technological mediators of the sense and perception of time. And in a sense, they made possible the very calculations which lay at the much later basis of measurement undertaken by the Galileos and Keplers of the early scientific era.

Turn now to a spatially mediated experience and note that the same invariants occur again. One of the most important technologies which allowed the science of the modern era to become truly experimental was optics. Lenses were developed in the tenth century, were already compounded by the thirteenth century, and simultaneously with the first explicit scientific observations, the microscope and telescope were invented.

Vision is embodied and mediated through lenses. What changes is what might be called a shift of focus from ordinary perception to the technologically mediated micro-dimension. Distance is reduced, what is far is brought near, but this is equivalent to saying that what was for ordinary vision a micro-feature is now made present. The microscope brings into view for the first time the small and unexpected creatures found in drinking water; the telescope reveals that the shaded areas of the moon are seas and mountains and craters. The span of space is changed, reduced, and the object is "brought closer." What was previously so distant as to be unperceived, is now perceived in a near-distance of optically mediated space. Again, both what is focal and what was the field of space changes under the transformations of technologically mediated perception.

This is to say, that through the use of technologies, experience had already become prepared for the scientific experience of the world. A world whose features could be considered as discrete units, a world whose micro-features would fascinate, a world conceived of under the sign of mechanical relations, was a world which was prepared for by the taken-for-granted technologically mediated experience of the Medieval Period.

Late Medieval experience of both time and space could be considered to be thoroughly embedded in and often mediated through technologies. One could expand upon these examples in many areas of life. One could also contrast these examples of technologically mediated perceptions of space and time with cultures which did not have clocks or lenses and note that time and space are differently perceived by the latter. But I shall now turn to a more dramatic example of the way experience and praxis are organized and examine a crucial case of long-distance spatial orientation, the variant development of a perceptual and technologically mediated perceptual navigational system.

Variant Long-Distance Spatial Orientation: Atlantic and Pacific Navigation

One of the features which stimulated the European development of technology, was the availability of ideas and devices from many areas of the world, an availability made possible through the early exploratory trips of Europeans. We are familiar with some of the historical events which were associated with this cross-cultural interchange: the Crusades, the travels of Marco Polo, the centuries of coastal voyages; and only much later, the full spice trade and voyages of conquest for gold and riches which fed the end of the Medieval Period. I shall focus here upon the development of cross-oceanic navigation as it contrasts with the Pacific variant.

Coastal navigation, essentially navigation within sight of land or never far from it, is distinctly different from transoceanic navigation. The principles or practice of coastal navigation and the body of knowledge which goes with it were known from ancient times. Such navigation was largely perceptual and traditional since observations of currents, animal life, noise and sight of breakers over shoal waters, wind patterns, etc., were necessary for safe coastal piloting. Fears of out-of-sight navigation were not merely those clothed with superstitions about the unknown (monsters, the end of the world, etc.) but were related to a lack of knowledge about how to return to a known area. In short, what was needed was a means of dependable spatial orientation across the expanse of uncharted ocean.

Early Western transoceanic navigation was successfully undertaken by the Vikings who traveled from Scandinavia, not only throughout Europe and the Near East in coastal raids, but also to Iceland, Greenland and Nova Scotia in the New World. How these voyages were undertaken lies somewhat obscured by a sparse historical record, except that we know that two features of navigation unique to Northern Europe were already known: a fixed star, the North Star (Polaris) was known and navigational calculations could be based upon this fixed point. And the primitive use of the lodestone which also points to a fixed area, was already common with the Vikings. Thus, although very simple, one can say that the very origin of transatlantic navigation was technological in a most primitive sense. Orientation was secured through a device.

If, however, one takes the voyages of Columbus as more typical, then the technological determination of orientation is abundantly clear. By 1492, the transition period for our purposes, not only is there a magnetic compass, but measured and careful cartography was known, and a more vast array of instrumentation was also available. Compass, astrolabe for calculating angles to the sun and other heavenly bodies, clocks (although not yet fully useful for ocean voyages) and various measuring devices were used for navigation. Columbus's daring voyage was a voyage undertaken through a technologically mediated orientation to possible space. (Columbus knew very well that the earth was round; that it was of approximately a certain size—although vastly underestimated by his era—and that it could be plotted through calculations via instruments.) His navigation already conceived of the world as a grid upon whose surface one moved, and his perceptions were instrumentally mediated. Thus our earliest voyages through the period of world exploration were voyages which were undertaken through technologies.

When one turns to the Pacific we find that the Polynesians and related peoples had, already 1000 years before the Vikings, explored and populated virtually every inhabitable island chain of a much larger ocean. Western explorers were amazed by the 200-foot-long catamaran war canoes which speedily navigated the Pacific, yet they did not pick up the secrets of Polynesian navigation at the time. One must conclude, on the basis of praxis, that both Atlantic and Pacific navigation were successful, but on examination, each was a distinct and different system.

Polynesian navigation was instrumentless; it operated without fixed points such as Polaris, which is not visible in the Southern hemisphere; nor did the Polynesians have the technological fixed point of the compass. It was a rather complex system of perceptual observations carried on through a secret tradition by a school of navigators.[4] I shall not outline all of the features of this perceptual system, but shall point to enough features to illustrate its subtlety:

(i) One key feature of the perceptual system was a highly developed sense of wave patterns. Waves march with regularity across the Pacific and the Polynesian navigators learned to use them for precise directional purposes. By judging the angle of swells in relation to the direction their canoes took, Polynesian navigators could maintain direction. They became so keenly aware of this wave harmonic that even when local storms confused the seas, they could detect the swell pattern engendered by the storm. (Often they would sit in the bottom of the canoe to feel this pattern—their claim was that only men could navigate so because they felt the pattern in their testicles.) They also were aware of what we would call refraction waves: swell patterns bend when they approach a land mass such as an island and the change in direction was detected and understood as an indication of a distant island.

(ii) Cloud and light patterns were also learned. Far over the horizon a column of cloud, slightly green tinted skies, and other more dense moisture indications would be read as the presence of an island. Again the indications were perceptual readings of the phenomena.

(iii) Although bird behavior and patterns were not unknown to European coastal navigators, the precision of observation which knew exactly how far each species strayed from land, the knowledge that a direction towards land could be obtained at dusk by returning birds, and even knowledge of which fish inhabited near-island waters enabled the Polynesian navigators to regard the ocean stretches as a familiar, readable world.

(iv) Star paths were learned and conveyed from generation to generation of navigators. Lacking an immovable pole star, the Polynesians developed a highly temporal, dynamic mode of reading star tracks over the horizon with changes of direction timed to moving locations. Indeed, all constants were in effect dynamic and temporally changing constants in this system.

Here was a navigational system which historically was at least equally successful in conquering transoceanic distances, a system which had more difficult tasks to perform since small island systems are harder to locate than continental masses, and a system which was thoroughly perceptual and historical. It was a system whose "map" of the earth was based upon perceptually acute readings of the ocean, without either a mathematics except for a time since (but no clocks) or an instrumentation. It was a variant orientational praxis.

One might very well expect that a variant praxis would be sedimented in a variant understanding of the world; and that certainly is the case. The Polynesian view was—if interpreted by Western standards—"animistic." The ocean was not perceived as either alien or strange, although its dangers and threats were clearly appreciated. It was a deity whose many natures could nevertheless be understood. It was the source of nurture and support and thus a voyage upon its face, while it may pose dangers, was not a voyage into the wild nor something over which humans could expect mastery.

Do not misunderstand the point I am making here: I am not claiming that this life-world is better than that of the technologically oriented Modern. But it is different. Its praxis, focused perceptually, achieves similar goals although it implicates a different understanding of the world. It is a world which does not become standing-reserve because the earth's bounties are conceived of differently.

One might also point out that the Polynesian world is one which is disappearing. Its navigational arts, though still extant among a small number of persons, have been replaced by the now highly micro-determined instrumented navigation of the West. Long voyages by islanders are now undertaken on trading schooners or ships. (Although their ability to sense land before the Westerner remains, trading schooner captains indicate that they have lapsed into only rough navigation because they know that their passengers will begin to sing when approaching their island, long before the Western captain knows it's near.) My point is that two differently patterned praxes implicate two different ways of understanding the world, and ours is and has been historically Technological for centuries, indeed virtually for at least a millennium.

If Heidegger is right, that the essence of Technology shows itself only recently, it is because we have failed to look at what was under our very noses for a long time. But Technology is like a set of spectacles: those who see through them and who have become accustomed to them, do not notice them. Thus that which is closest and most familiar to us, we have failed to notice. Yet what we have failed to notice turns out to be basic, perhaps the most basic thing about the very way in which we see the world.

Conclusion

I have suggested that there is a significant sense in which Technology is both historically and ontologically prior to science. This priority, I believe, is one which is not contrary to the more trivial sense in which the human use of technologies is both universal and archaic, common to all cultures whether or not they have developed science.

I have also suggested that the way in which this priority operates is at the level of a basic praxis within a life-world, a praxis which inclines or predisposes us towards what becomes a scientific world view. I have developed only some of its features—those which include a technologically mediated basic perceptual experience. This is an experience which harbors invariant characteristics such as transformed foci regarding ordinary and micro-dimensions of experience, a tendency towards discreteness and the atomization of things, and the enhancement of calculative activities. In this sense Technology at the level of familiar praxis precedes and sets the conditions for a science.

Science, in turn, becomes the coming to self-consciousness of these activities, a self-consciousness which both projects the form of life implicit in the praxis upon the universe, and a self-consciousness which becomes increasingly purified of diverse elements. Such a purification, however, is also a purification of the essence of Technology.

Even the Renaissance, enamoured of inventions, and its desire to measure and use the world, created its artifacts in the form of animal and human life. Da Fortana's measuring robots were conceived of in the form of fish, rabbits and birds. The predecessor of the steam boiler was the *sufflator*, literally "blower," whose shape was always that of a human head whose mouth blew forth the steam which powered various devices. Only gradually did the *abstraction* needed for contemporary Technology emerge, thus freeing technologies to be "scientific" as embodiments of a purely technological metaphysics.

The gradual movement to de-animate our technologies, to move towards purer *functionalism*, is both latent within technology and a preparation for a scientific world view. It is a long step from the symbolism of the clock whose movements represented the heavenly bodies to the bare, instantaneous numbers of the digital, but the movement is one towards a more totally technological and scientific representation.

There is one question still left unanswered in this chapter, the issue which separates idealist from materialist interpretations of science and technology. But it may begin to be understood in a different way, too. That issue is whether or not and in what sense *scientific* technology may be distinctly different from traditional technology. My answer is that in one sense it is different, in another not.

The sense in which it is not different is the sense in which technologies have and continue to have the same existential dimensions with respect to the humans who use them. Technologies may embody and mediate experience so that our life-world undergoes changes; technologies may be "other" than we as that to which we relate; and technologies may increasingly be surrounding features of our life-world. In each case these appearances of technology may be seen to be continuous with even the most archaic technology.[5]

The sense in which scientific technology differs from traditional technologies depends upon the synergistic interaction of a technology made abstract or purified through the self-consciousness connected with science. Thus the break from "natural" materials to the manipulation and creation of materials, the gestalts which occur between scientific fields, and the extrapolations made possible by revolutions in science could only happen when the essence of Technology has become manifest. But precisely because it has become so, we can now notice more distinctly and clearly that we are wearing eyeglasses, and we can begin to reflect upon the implications of that wearing.

References

Heidegger, Martin. 1962. *Being and Time*. New York: Harper and Row.

———. 1977. "The Question concerning Technology." In Basic Writings, translated by David Krell. New York: Harper and Row.

Ihde, Don. 1979. *Technics and Praxis*. Dordrecht: Reidel.

Mumford, Lewis. 1936. *Technics and Civilization*. New York: Harcourt, Brace and World.

White, Lynn, Jr. 1962. *Medieval Technology and Social Change*. Oxford: Oxford University Press.

———. 1972. "Cultural Climates and Technological Advance in the Middle Ages." *Viator* 2:171–202.

Chapter 13

Epistemology Engines
and the Camera Obscura

In November 2001, David Hockney's *Secret Knowledge: Rediscovering the Lost Techniques of the Old Masters* was published with great fanfare. It made the claim that many artists from the Renaissance on used a now antique technology, the *camera obscura*, to make their paintings. Hockney, always somewhat controversial, had hit an issue. The release was followed within a few weeks by an overflow crowd conference at New York University with historians and critics responding to this thesis—claimed to be radical by Hockney. The dominant response was—well—"*Heideggerian*." If Hockney is correct, then many revered artists could now be thought to have "cheated" by copying an image produced by a device, a sort of drawing-by-the-lines approach taken as "inauthentic" by the art crowd who implicitly favored hand and brush. Here is Heidegger:

> Human beings "act" through the hand; for the hand is, like the word, a distinguishing characteristic of humans. Only a being, such as the human, that "has" the word (mythos, logos) can and must "have hands." . . . The animal has no hands, nor are hands derived from paws, claws, or talons. . . . The hand has only emerged from and with the word. . . . [For Heidegger, this becomes the privileged place for hand in handwriting] because the word, as the essential region of the hand, is the essential ground of being human. . . . The word . . . symbolically inscribed . . . presented to vision is the written word, . . . script.

As script, however, the word is handwriting. [But, handwriting, in typical Heideggerian dystopianism, gives way to technology, the typewriter.] . . . It is not by chance that modern man writes "with" the typewriter. . . . This "history" of the kinds of writing is at the same time one of the major reasons for the increasing destruction of the word. The word no longer passes through the hand as it writes and acts authentically but through the mechanized pressure of the hand. . . . Mechanized writing deprives the hand of dignity in the realm of the written word and degrades the word to a mere means for the traffic of communication. (Heidegger, quoted in Heim 1987, 194–95)

Substitute the brush for the pen and you have much of the response to the Hockney thesis by many of the critics.

When this publicity hype happened, I and a number of friends, equally engaged in the histories of science and technology and instrumentation, had some mutual chuckles over e-mail, another doubtlessly "inauthentic" mode of communication.

Catherine Wilson, the eminent philosopher and historian of microscopes, had already forewarned me of the Hockney thesis prior to the book's appearance, after which we had a good laugh at the audacity of his proclaiming what most of the rest of us knew as common knowledge, i.e., that the Renaissance in both art and science was embodied through technologies, with the *camera obscura* being one favorite optical toy.[1] My prized 1929 *Encyclopedia Britannica*—the same one which contained Husserl's famous entry on phenomenology and Einstein's on relativity—had an extensive article on the uses of the *camera obscura* as early as Alberti's use in 1437. "The first practical step towards the development of the *camera obscura* seems to have been made by . . . Leon Battista Alberti, in 1437 . . . [referenced in] a biography, *Rerum Italicarum Scriptores*. . . . It is stated that he produced wonderfully painted pictures . . . [via] a small closed box through a very small aperture."[2] And most historians knew of Da Vinci's use of the *camera* to trace a cross from its image in 1450. And, what does one do about other technologies of drawing such as shown by Dürer and others (fig. 13.1)?

No, Hockney did not rediscover the secrets of the Renaissance, he simply republicized them. What may have been forgotten by some art critics and historians is how fully technologized the Renaissance and Early Modernity was. Might Galileo without his telescope be analogous to Caravaggio without his *camera*?

Figure 13.1. Man Drawing a Lute. Woodcut by Albrecht Dürer, 1525. Source: Wikimedia Commons: https://commons.wikimedia.org/wiki/File:D%C3%BCrer_-_Man_Drawing_a_Lute.jpg._Man_Drawing_a_Lute.jpg.

First, then, what is the *camera obscura*? In simple terms, it is a very large "pinhole camera"; its literal meaning is dark room. The optic effect which the *camera obscura* captures was known in antiquity—probably known to Euclid and the Hellenic Greeks, but first thoroughly described in the revival and innovative expansion of this knowledge by the Arabic natural philosopher, Al Hazen, in 1037. Al Hazen's *Optics* was the first major systematic treatise on light and optics and therein may be found a description of a *camera obscura*, which he probably used to observe a solar eclipse and which may have been the first anticipation of an early modern scientific use of the *camera* (fig. 13.2).

The earliest European references to optics did not occur until 1270, with essays by Erazmus Ciolek Witelo and Roger Bacon—but by the Renaissance, the *camera obscura* was clearly one of the common optical toys of the time. Here is how it works: A light source, "outside," is directly cast through a small hole and then is produced as an "image" on some flat surface "inside" the darkened room. Carefully now note the effects and the ways in which the *camera* transforms an "object" into an "image." The light

Figure 13.2. Camera Obscura. From Athanasius Kircher, *Ars magna lucis et umbrae*, 1646. Source: Wikipedia Commons: https://en.wikipedia.org/wiki/File:1646_Athanasius_Kircher_-_Camera_obscura.jpg.

source, here the sun, is produced in inverted form, and in two dimensions on the blank wall. (With the sun, these effects might not be noticed, but if the image cast is one of an indirectly lit up object, the inversion and the two-dimensionality is easily perceived.)

Our "Alberti" now can reproduce in a drawing this "automatically-reduced-to-Renaissance-perspective" image, emerge from the dark room and invert the drawing, and get the appreciative gasp of "verisimilitude" his audience noted. One can see from this simple example how thoroughly so-called Renaissance perspective is "technologically" produced. Its mathematization in one sense *follows* the optical effect.

Now having noted the role the *camera* and related technologies played in the development and refinement of practices such as Renaissance perspective, the new way of seeing which occurs via "technological mediations" illustrates the impact of optical technologies upon artistic practice. From this, we can equally note another indirect role played by the same technology. Al Hazen already noted in the eleventh century that the eye and the camera have certain analogous functions and shapes. Da Vinci later repeated this analogous observation: "When the images of illuminated bodies pass through the small hole into a dark room . . . you will see on the paper all those bodies in their natural shapes and colors, but they will appear upside down and smaller . . . *the same happens inside the pupil*" (Da Vinci, quoted in Bailey

1989, 66–67; italics mine). Here the *camera* begins to perform a *metaphoric* role. I say this because while there are some obvious analogical features: light through pupil, cast—as we now know much later than the Renaissance—upon the retina, etc., there are many others which are *disanalogous*. If the pupil is a "lens," it is a dynamic, variable one compared to the static one of the *camera*; the retina is a concave sphere in shape contrasting with the flat shape of the *camera*, etc., but metaphors emphasize similarities rather than dissimilarities.

To this point, the earlier role of the *camera* in its relation to art practices has been noted. Already familiar and in use in the sixteenth century, even greater familiarity continued into subsequent centuries. And, if first used in art practice, the *camera* was also adopted into science practice. Indeed, Al Hazen had already used it as an instrument for observing solar eclipses and in the seventeenth century the ever inventive Galileo used a modification of the *camera* conjoined to a telescope as a helioscope as the means of observing sun spots—one of his major discoveries (at least for Europe—Chinese knowledge of sun spot activity had already been recorded several centuries earlier). Here, however, rather than review the various optical devices and experiments associated with seventeenth century science, I want to take the lifeworld familiarity of such technologies in a more philosophic and epistemological direction.

One really large step in making the *camera* into a metaphoric device, into what I call an *epistemology engine*, occurs in the mid-seventeenth century. Both Descartes, in the *Dioptics*, and Locke, in the *Essay on Human Understanding*, specifically turn the *camera* into the very model for the production of knowledge, an *engine* of knowledge. Locke's description is the most isomorphic:

> External and internal sensation are the only passages I can find of knowledge to the understanding. They are . . . the windows by which light is let into this dark room; for methinks the understanding is not much unlike a closet shut from light, with only some little opening left, to let in external visible resemblances, or ideas of things without . . . [these] resemble the understanding of man, in reference to all objects of sight and the ideas of them. [His *tabula rasa* was, of course, the blank wall of the *camera*.] (Locke, quoted in Bailey 1989, 68)

Indeed, if we now use the *camera* as the model for knowledge production itself, one can get the outlines of both its terminology and its problems.

Point by point, the sun is "external" reality, *media res*: the image cast, the idea or "thinking" in the mind. The "subject," here invented as parallel to the "object," is *inside* the box of the body and sees only its own images/thoughts which correlate and represent that which is *outside*, or as it later becomes "external reality." But—and here comes problem one—if one's "subjective knowledge" is to be "true," it must *correspond*, like for like or isomorphically, with what is external. Thus Descartes plugs in his famous argument for God and his non-deceptive role guaranteeing the correspondence. One can also see here, at a glance, that the implied subject-as-homunculus also arises. What Descartes has done, is taken the metaphor of eye-*camera*, and expanded it into what I call an *epistemology engine* by making the metaphor stretch to: I=eye=camera.

But Descartes (and Locke) employ a "cheat code." That is, the full description of what occurs inside the box is paralleled by a description of what occurs *outside* the box, thus implicitly presuming a point of view—were Descartes to be right and not be himself limited to a description as if he were truly only inside the box—which he could not have were knowledge limited to being inside the box. In short, Descartes occupies a position that allows him simultaneously to see both inside the box and outside the box, the position Donna Haraway calls the "god trick." Descartes doesn't really need God since he has already opted the god position! It was thus a technological model that shaped the basic outlines of early modern epistemology. We humans modeled our own processes upon one of our own inventions, a technological artifact.

I could, and would like to go on, but the point I am making here is not one which engages in present debates about early modern epistemology—still vestigially virile in at least some disciplines today, although with the earlier Cartesian location of the implied "homunculus" inside the pineal gland, is now re-located inside the "brain"—it is instead intended to point to the way in which a technology plays an important role, one which even makes it into the very model of knowledge production itself.

To show this I want to take the *camera obscura* beyond both its Renaissance and seventeenth century settings and relate its later modifications to the same knowledge production activities. The *camera obscura* was a tool which through an amazing number of proliferations of technological development and modification can be shown to be one of the most versatile springs of scientific instrumentation, but it, by becoming a metaphor, also became the epistemology engine which justified and explained how scientific knowledge itself worked. And while this may in some ways fall into the Heideggerian observation that technology is *ontologically prior to science*, it is an observation which is much more concrete, empirical, and historical than Heidegger could be.

Nor can I here yield to the very strong temptation to show how early modern epistemology no longer works well, particularly in science, and with that follow moves to more contemporary equivalents of replacement epistemology engines—which I suspect all of you would recognize in the roles of computer and internet processes which are rapidly being metaphorized into cosmology itself—but instead I want to show how, in a much more concrete and phenomenological sense, the *camera obscura* invented science in different and recognizably actual ways:

1. The optical effect captured by the *camera* goes back to antiquity, probably known to Euclid, but first fully described by Al Hazen, who also seems to have used it to watch an eclipse—just as we do today with cardboard versions. Galileo also adapted a *camera* device, a helioscope attached to his telescope, which cast an image showing the sun to have spots—his report of this was the first publication to get him into trouble with the Vatican.

2. But, its first major social use was in art, as indicated, and the camera was a sort of "automatic" process by which "Renaissance perspective" was produced. The *camera* "automatically" reduced three-dimensional objects to two-dimensional projections, with "perfect" proportionality. Note that this is a transformation of a "natural object" into what we often call an "image." If one extends this insight, one can see that the way of seeing associated with Renaissance perspective, mathematized as "geometrical projections," invents a "geometrical method" which precedes by a couple of centuries what is usually taken as the birth of early modern science. I also want to note that this early use of the *camera* is limited to its *isomorphic imaging* possibilities, i.e., to make "realistic" depictions. If the style of vision that is associated with "Renaissance perspective" is one of "realism," it owes much of this to *camera production*, and in this Hockney is justified.

3. This same isomorphic possibility—as I am sure you have already anticipated—becomes, in the nineteenth century, another "camera," this time a *photographic one*. There is no example in the history of technology that I know to have exceeded the speed of dissemination than that occupied by

photography. The process, actually invented by Joseph Niépce in 1826, was bought out by Louis-Jacques-Mandé Daguerre and publicized in his best seller on how to build a camera in 1839. Scientists began using this process almost immediately, as in this first shot of the gibbous moon in 1840. Only two years later (1842) the astronomer, J. W. Draper, photographed a solar spectrum. And while one can easily see that the photographic camera is a reincarnation, a simple modification of the earlier *obscura*, one should not pass by without noting its different *technological transformations* of objects into images. Early photography was *still*, it "fixed" its images, and for science this was important because the fixed image could be returned to again and again for deeper and deeper analysis. Note, too, in early form, how the technology non-neutrally produced instrumentally *selected* objects. Exposure time was such that at first only still objects—landscapes and architecture mostly, could be photographically imaged. That also partly accounts for the "stiff poses" of early portraiture. (I cannot follow here the proliferation of other possibilities that photography took up with faster shutter speeds and the like, but my general point is that photographic camera technology, like all technologies, is non-neutrally selective and *transforms* all its objects into what we now usually call images.)

4. A third reincarnation of the *camera obscura* might be missed, because what counts for imaging in this version takes a very different shape, a *non-isomorphic shape*. I refer to *spectroscopy*, which got its start via Isaac Newton. Newton, like Galileo, an actual optics maker, wrote in 1666, "I procured me a triangular glass prism to try therewith the celebrated phenomenon of colors. . . . Having darkened my chamber and made a small hole in my windowshuts to let in a convenient quantity of the sun's light, I placed my prism at its entrance, that it might be thereby refracted to the opposite wall" (Isaac Newton, letters, quoted in Bolles 1997, 184f.). In short, another *camera obscura*, simply now replacing the hole—which very often was also lensed—with a prism. The result was that the image cast was that of the rainbow spectrum. Newton himself did not go on to invent the spectroscope—but from these experiments

he did derive a theory of color and brilliantly found a way to overcome the color aberrations associated with any refractive telescope over 30 power—Newton invented the reflecting telescope which re-focuses the different frequencies of colored light into a single location by way of a parabolic mirror. It was not until more than a century later that chemists, still playing with the prism version of the *camera*, found that by replacing the round hole of the classical *camera* with a slit, they could get a far better image, one which was found to have dark lines rather than fused colors; these are called Fraunhofer lines after a chemist by that name produced them in 1814. These were, again later, finally discovered to be unique chemical signatures such that one could "read" the chemical signatures of the sun and stars. Today, spectroscopy has proliferated into thousands of types of instruments, which in a parallel with optical tele-scopy, now can exceed the ranges of light itself. Spectroscopy allows us to "read" nature's bar codes. But the instruments are recognizable variations upon the *camera obscura*.

5. Nor does the story end there. My colleague, Bob Crease writes for *Physics World* and he recently polled physicists about what they took as the ten most beautiful experiments in the history of physics, and the *Science Times*, September 24, did an article about these. Out of the ten, three are about light and all are variations upon *camera obscura*–like instruments. Newton's prism version is listed and ranks number 4. A little over a hundred years later, Thomas Young (1803) makes another modification on the *camera*, instead of a lensed hole or a diffraction slot, he uses a *double slit*.

> "He cut a hole in a window shutter, covered it with a thick piece of paper punctuated with a tiny pinhole and used a mirror to divert the thin beam that came shining through. Then he took a "slip of a card" . . . and held it edgewise in the path of the beam, dividing it in two. . . . The demonstration was often repeated over the years using a card with two holes to divide the beam . . . [these are now called] double-slit experiments" (Johnson 2002, 3).

And, voilà, yet another *camera* variant. Physicists ranked Young's experiment number 5. Rank number 1 stretches the double slit a bit, but only because instead of light, a beam of electrons is passed through a double-slit device yielding the famous quantum "wavicle" or particle/wave phenomenon so dear to contemporary physics—the actual experiment occurred in 1961 by Claus Jonnsen at Tubingen. The *camera* is brought into the late twentieth century—and I sometimes wonder what a triple slit might do?

At the very least, these variations, or technological trajectories from the early, basic *camera obscura*, allow the sciences to see, "read," and probe into the microscopic and macroscopic phenomena, which expressed, are the knowledges produced by science and its menagerie of *camera obscurae*. These are the productions that "invent" much modern science.

I want, however, to leave you not simply with this epistemic history as it were, but to suggest something more provocative: Each variant upon the *camera*, isomorphic, non-isomorphic, and today often *constructive*, produces visual displays, images, which are what are seen and read by the trained scientific hermeneut. But *the scientific object is, precisely in its strongest sense, that transformed phenomenon presented in and through the instrumental technology. Nature does not come to modern science raw or naked, but technologically transformed. The natural object must be transformed into the "scientific object" to be observed. That is the deeper secret of the camera obscura.* Science, however "realist," or however "socially constructed," is also *technologically constructed.*

References

Bailey, F. Lee. 1989. "Skull's Darkroom: The *Camera Obscura* and Subjectivity." In *Philosophy of Technology*, edited by Paul T. Durbin. Dordrecht: Kluwer.

Bolles, Edmund Blair. 1997. "Isaac Newton and Robert Hooke, Dispute on the Nature of Light." In *Galileo's Commandment*, edited by Edmund Blair Bolles, 184–92. New York: Freeman.

Encyclopedia Britannica. 1929. Vol. 4. New York: Encyclopedia Britannica.

Heim, Michael. 1987. *Electric Language: A Philosophical Study of Word Processing.* New Haven, CT: Yale University Press.

Hockney, David. 2001. *Secret Knowledge: Rediscovering the Lost Techniques of the Old Masters.* New York: Putnam.

Johnson, George. 2002. "Here They Are, Science's 10 Most Beautiful Experiments." *New York Times*, September 24.

Pollack, Peter. 1977. *The Picture History of Photography*. New York: Abrams.

Wilson, Catherine. 1995. *The Invisible World: Early Modern Philosophy and the Invention of the Microscope*. Princeton, NJ: Princeton University Press.

Chapter 14

The Phenomenology of Scientific Imaging

If you are convinced by the narrative history just traced, then the expansion of hermeneutics, already convergent toward the sciences, should be ready to *enter* into the realm of the sciences and their "thingly" concentrations. Premodern hermeneutics naively believed that Nature contained the "writing" placed there by God in the creation. But we cannot return to that context or time. I shall use here a different and constructed metaphor to replace the ancient one. The postmodern hermeneutics of things must find ways to give *voices* to the things, to let them *speak from themselves.* The source of this metaphor is phenomenological and draws from my own research history. For both Husserl and Merleau-Ponty, voiced language is a *bodily* and a fully perceptual activity. It is that materiality of perceptual activity which I seek in this thingly hermeneutic.

Nearly three decades ago, in my own research history, I began investigating *auditory* perception, which plays such important human roles in speech and listening (and music), roles which I thought the tradition had often neglected and which thus sometimes led to the impoverishment of philosophical richness in understanding human being-in-the-world (see Ihde 1976). One of the lessons which emerged from these investigations was that so many things which are simply seen or viewed appear *silent.* Yet everything potentially *has a voice* under one of two conditions: first, seemingly silent things can be given voices—that has frequently been the musician's gift to our experiences. Even stones, struck, "speak" forth. Percussion is one way of giving voices to things.

The voice which is given to things, or elicited bodily from things, however, is very *complex*. First, if the thing is struck on the model of musical percussion, the voice is not single but is a duet. The sound produced is both the voice of the thing struck and the voice of the striking instrument. Substituting a bell for the stone example, a bell struck with a wooden mallet produces a different "duet" of sound than one struck with a brass mallet, or again with a rubber mallet. Here there is some implicit *science* as well. The relativistic and quantum sciences of the late twentieth century are self-consciously aware of this instrument-object interface. Even to measure the temperature of a pot of water blends together the eventual temperatures of the water and of the measuring thermometer. As we shall see later, this becomes one practical reason for introducing what I call "instrumental phenomenological variations" into our letting things speak, that is, the use of multivariant instrumental measurements.

The second condition is one which can appear only at the microlevel. Often a thing may already be "sounding" below or above the levels of our "earhole" perceptions (my substitute here for the common use of the "eyeball" observational). The radiations of wave phenomena—microwaves above or below earhole perceivability—may again become presentable to our hearing *if mediated by the proper instrument*. Thus, for a second time, the material intervention of a material artifact, a *technology*, enters into the conditions for giving things a voice, or allowing the thing to "speak" for itself.

As I am writing this, "Pathfinder" and its little robot, "Sojourner" (named for Sojourner Truth), are exploring rocks on Mars. The examples from metaphorical auditory experience are obviously apt for this set of events. The Sojourner, equipped with an alpha proton X-ray spectrometer, must approach each rock carefully, touch it, and bombard the rock with X-rays, thereby probing the ions in the rock by radiative "percussion"—the "giving voice" metaphor, here at the micro level, is appropriate to the high-tech, engineering-embodied space science of the present. The ions activated "bespeak" the chemical composition of the rock.

What, however, is "hermeneutic" in all this? To answer that is the real task of this chapter. Following in a slightly different way the suggestions of Rouse previously noted, I want to differentiate between what I shall call a "weak" and a "strong" hermeneutic program with respect to science. The "weak" program is an attempt to reconstruct accounts of science praxis, showing the implicit *hermeneutic practices* already at play within science. This amounts to a claim that in one of its knowledge-producing dimensions, science is already a hermeneutic practice. Here the task is to show that various interpretive activities within science practice are already hermeneutic in form.

The "strong" program is potentially more normative. It will be an attempt to push, positively, certain P-H practices by way of suggestion and adaptation toward science practice. Thus, I will be outlining a more aggressive hermeneuticizing of science, although based upon forefront research fields as now emerging in late Modern science.

Modifying the P-H Tradition

The first step in expanding hermeneutics into the thingly sciences is to *modify* the phenomenological-hermeneutic tradition itself, clearing it of some of its own last prejudices about science, but from within some of its own potential insights. The thrust of this modification is one which reconstructs the understanding of science and the lifeworld. As I argued in several of the previous chapters, most explicitly in chapter 4 [of *Expanding Hermeneutics*], classical phenomenology—particularly with Husserl of the *Crisis* and "The Origin of Geometry,"[1] and with Merleau-Ponty in *Phenomenology of Perception*[2]—tended to interpret science as separated off from the basic perceptual-bodily activity of the lifeworld.

This separation, however, comes about in part because the construal of science in early phenomenology was in keeping with the P-H Binary I have alluded to, the binary which tended to view science as a propositional and theory-biased special activity, rather than the institutionalized, embodied, and *technologically mediated* science which late Modern philosophers have more often taken it to be. In short, I am arguing that this science, *technoscience*, has never been separated off from the lifeworld but is a unique Modern, now-postmodern, activity which produces its knowledge through its very technological embodiment.[3]

Of course, this means that the P-H tradition must accept within itself the mediated forms of intentionality which come through technologically mediated experiences, alongside and with all other bodily perceptual activities. Science, in this view, is one highly developed, complex, and skilled mode of mediated praxis, but in principle it is not out of step with Heidegger's "hammer" or Merleau-Ponty's "feather."

To accomplish this shift within P-H understanding of science, I shall revert to a Galilean parable—a revisitation to a fictional, phenomenologized Galileo as Godfather of modern science. The historical Galileo, while "metaphysically" proclaiming a reduced and abstract world of inertial motion of material objects, in practice employed an array of instruments through which his early science made its discoveries. While not the inventor of the compound lens telescope, Galileo developed and perfected over a hundred of these instruments. He became convinced that telescopically mediated perception was "better" than eyeball perception, and one of his arguments involved evidence that a certain "halo" around celestial objects could be seen with his telescopes which could not be detected by naked eye (Brown 1985, 487). The irony is, of course, that this effect was what we know as an "instrumental artifact," an effect of the technology, not of the referent object.

We also know, historically, that Galileo proclaimed that "any man" could see what neither the ancient philosophers nor the church fathers could see—but, only under the condition that Galileo "teach" the process of telescopic seeing (Brown 1985, 489). The initial perceptual ambiguities of planetary features and satellites were, again, associated with the primitive technology of Galilean optics. This stage of early, often hard to resolve object referents, only later resolved by better instrumentation, has long been a feature of the history of science.[4] The handheld telescopes of Galileo were difficult to use and, particularly, to "fix" upon the referent objects. Beginning astronomers today can note this same problem set when viewing through handheld binoculars, for example.

Third, the historical Galileo clearly had his fascination centered upon the celestial objects, on what was "out there," and gave these observations his primary attention. This, again, is a typical feature found associated with the uses of new technologies. Its echo today can be found in the hype accorded to all sorts of technologies, particularly computers. There is a tendency to overestimate what can be gained with many novel technologies.

But, now, I want to reinvent Galileo, in a more thoroughly phenomenological style. He has, for some time now, been tinkering with his optical instruments and has already noted the surface features of the biggest nighttime object, the Moon. When ancient theory called for the Moon to be smooth surfaced, purely circular, and of a celestial perfection not possible upon Earth, the very first sightings had to have been dramatic. Even Galileo's instruments showed—immediately and dramatically—craters, valleys, mountains, plains, shadowed and lighted irregularly with respect to the phase of the Moon vis-á-vis its light source, the Sun. One cannot blame him for the initial enthusiasm that this first discovery of an "instrumentally mediated *realism*" which the telescope delivered. Nor could one avoid the later sense which would maintain itself into the contemporary world, of "effects which will simply not go away."[5]

We now, however, need to *phenomenologize* this fictional Galileo. He begins his description of the phenomenon, at first noematically, that is, with a concern for the "out there" object:

Spatiality has just undergone a set of dramatic changes; suddenly the Moon has mountains, craters, and so on, which mean that what was previously more "distant" is now "closer." But what makes it closer, and what changes occur? First, the "closer" Moon (through the telescope) has now displaced its previous context. It no longer occupies its relatively located and smaller appearance within the overarching heavens. In relation to its previous field, it has radically changed. Magnified, the Moon is "closer."

Did the Moon simply change its distance? Phenomenologically, every noematic shift is correlated with a noetic shift, that is, with a shift of the positioned perspective which is refracted from every visual observation *back to the position of the embodied observer.* This new, instrumental distance is thus a difference in distance between my body and the Moon (within the mediation). Indeed, it is equivalent to say that the Moon is now "closer" to me, or I am "closer" to the Moon—it is the *relational, or relativistic, distance* which has changed.

This phenomenon, historically, became known as "apparent distance." Why apparent? One possible answer is that the "real" distance between Galileo and the Moon did not change: it remained 240,000 miles distant whether or not he saw it through the telescope. This, of course, either privileges the measurable distance between the actual body of the observer and the referent object (thus, implicitly, does this mean that the eyeball observation is privileged as well?) or is constructed by some variation upon Cartesian-Newtonian space which presumes—not, as it claims, a nonpositional measurement—a measuring stance from an ideal or god's eye perspective which simultaneously sees both Galileo and the Moon.

I have described this version of "apparent" versus "real" distance in this way to show that there is lurking here a *variation* between eyeball and instrument-mediated perceptions. If the reflexivity of intentionality is maintained, it is the entire Gestalt which changes between direct and mediated perceptions. The relative distance, telescopically, is reduced. And the bringing near of the Moon occurs within the new context.

Phenomenologically, the relational distance is the *intentionality distance* which must include both referent object and perceiving, perspectival "lived" body, but not in the same way as in Cartesian-Newtonian frames. This phenomenological measurement must be reflexive and must utilize means which determine the (apparent) distance from within the correlation. What must be avoided is the ideal observer or god's eye simultaneous sight. Instead, were I to invent a measuring technology, it could be something like a radar probe in which the known speed of the signal is timed against its return. This space-time method is reflexive and avoids the Cartesian-Newtonian implied external perspective.

The new experience also implicates the very sense of one's body as well. In the beginner's experience, this is easily detectable when first trying to "fix" the celestial object. One can notice the "wavering" of the object, but this wavering is simultaneously the wavering of my holding of the instrument. The Moon's magnified character is simultaneously the *reflexive magnification of my bodily motion.* Both the Moon and I have changed, and both areas are possible areas for further and deeper investigation. If I didn't know the Moon had mountains (before instrumental mediation), neither did I know that I constantly had micromotions to my bodily position when actively viewing something other than myself. But both noemtic and noetic transformations are equally detectable in this new experience.

But, again within the context of changed Gestalts, there is a detectable difference in one's sense of body. The now suddenly micromotional body experience also conveys a sense of "irreality" when compared to my usual, already familiar actional sense of motion upon the Earth. A small indication of this occurs whenever anyone gets a new prescription for eyeglasses: one has to relearn, in however minuscule ways, the distancing between the ground and walking, and so on. Once learned, the "irreality" either is diminished or virtually disappears as the instrument is properly "embodied" into the new (now normalized) experience. Our phenomenological Galileo already knows how to use the telescope and can teach others to use it by applying his own experience.

The historical Galileo's interest, however, remained outwardly, externally oriented. And that interest can become—became—a *technological trajectory* which could be refined and followed. To enhance "fixing" the celestial object, first tripods, later more complicated machinery built into the now optical system, which includes motorized devices to keep the telescope fixed upon the section of the Moon being studied, was the trajectory historically followed.

These had to entail growing knowledge of Earthly and heavenly motions, to bring to a halt for observational purposes what was always in relative and now magnified motion into the technological development. Second, these developments which help to "fix" or stabilize observations without bodily training are also developments which can and do "distance" the previously direct bodily activities from the original contexts. The actual—relativistic—bodily seeing is replaced by the technoconstruction which allows the vision to be mechanically stabilized. The Greek preference for the eternally fixed remains within the trajectory being followed. Machines can "embody" metaphysics.

A trajectory is, however, a *choice*. In the original Galilean situation, there was multistability, and with it a number of possible choices. For example, had Galileo been fascinated with the revelation of his bodily micromotion discovered through use of the telescope, he might have taken a different direction. And later investigators did follow that trajectory. Contemporary empirical psychology employs a range of instruments which focus precisely upon micromotions of the human body. Microtiming of response time, minuscule eye motion, reaction time as a factor lying between autonomic and voluntary responses, and so on all were implicit from the instrumental beginnings, but historically came much later in the histories of the sciences.

This is to say that the historical Galileo *chose* certain, rather than other, interests, followed one, rather than multiple, instrumental trajectories, and in the process, insofar as he was followed by a community of like-minded fellows, opened *a* pathway for modern science which early favored astronomy, physics, and geometry.

Before leaving our reinvented Galileo, I want to follow one more line of phenomenological inquiry with respect to the *lifeworld*. Wanting to observe several of the dramatic comets which became visible in the last few years, I purchased a high-quality reflecting telescope, which my family, visitors, and I have enjoyed in the clear summer nights in Vermont. The Moon, the easiest target in the sky, is also the easiest to use to demonstrate "instrumental realism." A puzzle, however, has begun to emerge: when we look at the Moon with the naked eye, it always appears to have light and dark areas (the old "Man in the Moon" phenomenon), such that my son and I began to wonder how the ancients could ever have thought the Moon to have a pure, mirror-like, feature less surface. We simply *cannot* see it that way. But has our seeing already become post-telescopic seeing, and the Moon has changed, not just under the actual conditions of telescopic viewing, and now "contains" within the eyeball example the "residue" of the telescopic?[6]

One even more dramatic such effect occurred in 1996. We had been viewing Jupiter, whose four largest moons are dramatically visible through the telescope. These moons move in their orbits and thus clearly and quickly change position from night to night—another easy case of instrumentally delivered "realism." Then, one clear night, driving home from a nearby concert, I happened to glance up and saw—with the naked eye—Jupiter "with horns," or a clearly visible line out to each side of the disc. And I immediately recalled that Sandra Harding refers to Dogon (African) observations in the eleventh to thirteenth centuries which refer to Jupiter in just such a way.[7] I had simply rediscovered an ancient observation but, having done so, now have a "different" Jupiter within both eyeball and telescopic vision. Like the textured Moon, Jupiter no longer can revert to simple disc. This is, in short, a lifeworld accretion, which follows an irreversible direction. While there are different Gestalts for naked and mediated perceptions, there is also an interaction and overlap which through familiar embodiment shapes the contemporary texture of the lifeworld.

This revisitation of Galileo has been intended to show something concretely about a P-H-revised understanding of science within the lifeworld. From here, the expansion of hermeneutics, first as elicited from actual practices of science and then in to extrapolations of a more deliberate sort, can be done.

The "Weak Program": Hermeneutics Implicit within Science

In its broadest (and earliest) sense, hermeneutics could be rendered "interpretive activity." The metaphors of a more *linguistic* sort have tended to dominate, from the "Book of Nature" to my own "giving voice" regarding things. Add postmodernist emphases upon "textuality," and we note that much of interpretive activity has fallen under linguistic metaphors.

This is often the case even within the sciences, particularly if linguistics includes its *semiotic* dimensions. It is apparent in a transparent way within contemporary genetic sciences, where one now has strings of "codes" which can be determined and deciphered: genes "express" themselves in a technical sense which is not semiotic, but which carries the vestige of linguistic overtones. The same may be said of computer science with its binary coding and of digital processes in which "bytes," "pixels," and other "codes" are invoked, again carrying the linguistic overtones noted. One can even add the intermixture of genetic and computer semiotics as not only roughly parallel, but probably arising from a more or less implicit *Zeitgeist* which favors this set of grand metaphors. These metaphors both are more sophisticated than the earlier dominant "mechanical" metaphors and are able to carry the greater degrees of *complexity* which contemporary science demands. The ancient mechanical clock now seems to be replaced by the digital computer as the instrument of choice for our metaphors. If this is a trace of hermeneutics in science—and I think it is—it remains an accidental, rather than a deliberately constructed, one.

The vector I have been taking, however, is a somewhat different one. I have been arguing for and demonstrating a much more *bodily* and *perceptualistic* mode of interpretive activity. This moves somewhat tangentially to the dominance of the linguistic, and substitute features from perceptual experience which I am holding carry an appropriateness for the *thingly* which I take not to be somehow reducible to the secretly "linguistic."

I am tempted here—although I realize the dangers—to note that perceivability is an action which overlaps humans and animals probably more than language activity does. Of course, qualifications must immediately be entered: our perceivability is intertwined with linguistic and cultural interconnections which are distinctly human. That is because our "worlds" are also unique insofar as we are seldom prey who are actually eaten, when compared to our cousins, both domestic and wild, which frequently find this pattern in their lifeworld. The indicators, though, become somewhat more dramatic in the cases in which we try to enter, penetrate, or translate animal "languages" or teach them variants of our own. At best, there are suggestive intimations with chimpanzees or dolphins, and the specter of Quinean intranslatability is much more problematic with these quasi-languages than in the case with natural languages among humans—even among philosophers! Yet we might suspect that perceivability—of fast-moving objects, of oncoming objects, of changes in atmosphere—contains a greater overlap between us and our cousins.

If this is so, please note that I am not therefore arguing that perceivability is a "lower" or more "primitive" action than linguistic action as such. Quite to the contrary, I am arguing parallel to many of Dreyfus's observations that perceptual-bodily activity is both basis and implicated with all *intelligent* behavior (see Dreyfus 1993, ch. 7). And, if my broader overlap speculation is valid, then for an interpretive activity with the thingly, we need something more than "textuality."

Much of the line I have argued with respect to the history and philosophy of science is that Modern to late Modern science is what it is because it has found ways to enhance, magnify, and *modify* its perceptions. Science, as Kuhn and others after him seem to emphasize, is a way of seeing. Given its explicit late Modern hyper-*visualism*, this is more than mere metaphor.

There remains, deep within science, a belief that seeing is believing. The question is one of how one can see. And the answer is One sees *through, with, and by means of instruments.* It is, first, this perceptualistic hermeneutics that I explore in the weak program.

Scientific Visualism

It has frequently been noted that scientific "seeing" is highly visualistic. This is, in part, because of historical origins, again arising in early Modern times in the Renaissance. Leonardo da Vinci played an important bridge role here, with the invention of what can be called the "engineering paradigm" of vision.[8] His depictions of human anatomy, particularly those of autopsies which display musculature, organs, tendons, and the like—"exploded" to show parts and interrelationships—were identical with the same style when he depicted imagined machines in his technical diaries. In short, his was not only a way of seeing which anticipated modern anatomies (later copied and improved upon by Vesalius) and modern draftmanship, but an approach which thus visualized both exteriors and interiors (the exploded style). Leonardo was a "handcraft imagist."

The move, first to an almost exclusively visualist emphasis, and second to a kind of "analytic" depiction, was faster to occur in some sciences than in others. In astronomy, analytic drawing of telescopic sightings was accurate early on and is being rediscovered as such today. The "red spot" on Jupiter was already depicted in the seventeenth century. But here, visual observations and depictions were almost the only sensory dimension which could be utilized. Celestial phenomena were at first open only to visual inspection, at most magnified through optical instrumentation. It would be much later—the middle of the twentieth century—that astronomy would expand beyond the optical and reach beyond the Earth with instruments other than optical ones.

Medicine, by the time of Vesalius, shifted its earlier tactile and even olfactory observations in autopsy to the visualizations á la da Vincian style, but continued to use diagnostics which included palpations, oscultations, and other tactile, kinesthetic, and olfactory observations. In the medical sciences, the shift to the predominantly visual mode for analysis began much later. The invention of both photography and X-rays in the nineteenth century helped these sciences become more like their other natural science peers.

Hermeneutically, in the perceptualist style of interpretation emphasized here—the progress of "hermeneutic sensory translation devices" as they might be called—*imaging technologies* have become dominantly visualist.

These devices make nonvisual sources into visual ones. This, through new visual probes of interiors, from X-rays, to MRI scans, to ultrasound (in visual form) and PET processes, has allowed medical science to deal with bodies become transparent (Kevles 1997).

More abstract and semiotic-like visualizations also are part of science's sight. Graphs, oscillographic, spectrographic, and other uses of visual hermeneutic devices give Latour reason to claim that such instrumentation is simply a complex *inscription-making device* for a visualizable result. This vector toward forms of "writing" is related to, but different from, the various isomorphic depictions of imaging. I shall follow this development in more detail later.

While all this instrumentation designed to turn all phenomena into visualizable form for a "reading" illustrates what I take to be one of science's deeply entrenched "hermeneutic practices," it also poses something of a problem and a tension for a stricter phenomenological understanding of perception.

Although I shall outline a more complete notion of perception below, here I want to underline the features of perception which are the source of a possible tension with scientific "seeing" as just described. Full human perception, following Merleau-Ponty, is always *multidimensioned* and *synesthetic*. In short, we *never just see something* but always experience it within the complex of sensory fields. Thus the "reduction" of perception to a monodimension—the visual—is already an abstraction from the lived experience of active perception within a world.

Does this visualizing practice within science thus reopen the way to a division of science from the lifeworld? Does it make of science an essentially reductive practice? I shall argue against this by way of attempting to show that visualization in the scientific sense is a deeply *hermeneutic practice* which plays a special role. Latour's insight that experiments deliver *inscriptions* helps suggest the hermeneutic analogy, which works well here. Writing is language through "technology" in that written language is inscribed by some technologically embodied means. I am suggesting that the sophisticated ways in which science *visualizes* its phenomena is another mode by which understanding or interpretive activity is embodied. Whether the technologies are translation technologies (transforming nonvisual dimensions into visual ones), or more isomorphically visual from the outset, the visualization processes through technologies are science's particular hermeneutic means.

First, what are the epistemological advantages of visualization? The traditional answer, often given within science as well, is that vision is the "clearest" of the senses, that it delivers greater distinctions and clarities, and this seems to fit into the histories of perception tracing all the way back to the Greeks. But this is simply *wrong*. My own earlier researches into auditory phenomena showed that even measurable on physiological bases, hearing delivers within its dimension distinctions and clarities which equal and in some cases exceed those of visual acuity. To reach such levels of acuity, however, skilled practices must be followed. Musicians can detect minute differences in tone, microtones, or quarter tones such as are common in Indian music; those with perfect pitch abilities detect variations in tone as small as any visual distinction between colors. In the early days of auditory instruments, such as stethoscopes, or in the early use of sonar, before it became visually translated, skilled operators could detect and recognize exceedingly faint phenomena, as clearly and as distinctly as through visual operations. Even within olfactory perception, humans—admittedly much poorer than many of their animal cousins—can nevertheless detect smells when only a few molecules among millions in the gas mixture present occur in the atmosphere. In the realms of connoisseurship such as wine tasting, tea tasting, perfume smelling, and the like, specifics such as source, year, and blend—even down to individual ingredients—can be known. It is simply a cultural prejudice to hold that vision is ipso facto the "best" sense.

I argue, rather, that what gives scientific visualization an advantage are its *repeatable Gestalt features* which occur within a technologically produced visible form, and which lead to the rise and importance of *imaging* in both its ordinary visual and specific hermeneutic visual displays. And, here, a phenomenological understanding of perception can actually enhance the hermeneutic process which defines this science practice.

Let us begin with one of the simplest of these Gestalt features, the appearance of a figure against a ground. Presented with a visual display, humans can "pick out" some feature which, once chosen, is seen against the variable constant of a field or ground. It is not the "object" which presents this figure itself—rather, it is in the interaction of visual intentionality that a figure can appear against a ground.

In astronomy, for example, sighting comets is one such activity. Whether sighted with the naked eye, telescopic observation, or tertiary observations of telescopic photographs, the sighting of a comet comes about by noting the movement of a single object against a field which remains relatively more constant. Here is a determined and trained figure/ground perceptual activity. This is also an *interest-determined* figure/ground observation. While, empirically, a comet may be accidently discovered, to recognize it as a comet is to have sedimented a great deal of previous informed perception.

These phenomenological features of comet discovery stand out by noting that the very structure of figure/ground is not something simply "given" but is *constituted* by its context and field of significations. To vary our set of observables, one could have "fixed" upon any single (or small group) of stars and attended to these instead. Figures "stand out" relative to interest, attention, and even history of perceivability *which includes cultural or macroperceptual features* as well. For example, I have previously referred (see Ihde 1986, 128–29) to a famous case of figure/ground reversibility in the history of aesthetics. In certain styles of Asian painting, it is the background, the openness of space, which is the figure or intended object, whereas the almost abstract tracing of a cherry blossom or a sparrow on a branch in the foreground is now the "background" feature which makes space "stand out."

When one adds to this mix the variability and changeability of instruments or technologies, the process can rapidly change. As Kuhn has pointed out, with increased magnifications in later Modern telescopes, there was an explosion of planet discoveries due to the availability of detectable "disc size," which differentiated planets from stars much more easily (Kuhn 1962, 115–16).

I have noted in the previous section that Latour, in effect, sees instruments as "hermeneutic devices." They are means by which *inscriptions* are produced, visualizable results. This insight meshes very nicely with a hermeneutic reconstrual of science in several ways.

If laboratories (and other controlled observational practices) are where one prepares inscriptions, they are also the place where objects are made "scientific," or, in this context, *made readable.* Things, the ultimate referential objects of science, are never just naively or simply observed or taken, they must be *prepared* or *constituted.* And, in late Modern science, this constitutive process is increasingly pervaded by technologies.

But, I shall also argue that the results are often not so much "textlike," but are more like repeatable, variable *perceptual Gestalts.* These are sometimes called "images" or even pictures, but because of the vestigial remains of modernist epistemology, I shall call them *depictions.* This occurs with increasing sophistication in the realm of *imaging technologies* which often dominate contemporary scientific hermeneutics.

To produce the best results, the now technoconstituted objects need to stand forth with the greatest possible clarity and within a context of variability and repeatability. For this to occur, the conditions of instrumental transparency need to be enhanced as well. This is to say that the instrumentation, in operation, must "withdraw" or itself become transparent so the thing may stand out (with chosen or multiple features). The means by which the depiction becomes "clear" is constituted by the "absence" or invisibility of the instrumentation.

Of course, the instrumentation can never *totally* disappear. Its "echo effect" will always remain within the mediation. The mallet (brass, wood, or rubber) makes a difference in the sound produced. In part, this becomes a reason in late Modern science for the deliberate introduction of *multivariant* instrumentation or measurements. These *instrumental phenomenological variations* as I have called them also function as a kind or multiperspectival equivalent in scientific vision (which drives it, not unlike other cultural practices, toward a more postmodern visual model).

All of this regularly occurs within science practice, and I am arguing that it functions as a kind of perceptual hermeneutics already extant in those practices. I now want to trace out a few concrete examples, focused upon roles within imaging technologies, which illustrate this hermeneutic style.

Galileo's hand-held telescopes undertook "real time" observations, with all the limitations of a small focal field, the wobbliness of manual control, and the other difficulties noted above. And, while early astronomers also developed drawings—often of quite high quality—of such phenomena as planetary satellites, the isomorphism of the observation with its imaged production remained limited.

If, on the other hand, it is the repeatability of the Gestalt phenomenon which particularly makes instrumentally produced results valuable for scientific vision, then the much later invention of *photography* can be seen as a genuine technological breakthrough. Technologies as perception-transforming devices not only magnify (and reduce) referent phenomena, but often radically change parameters either barely noted, or not noted at all.

It would be interesting to trace the development of the camera and photography with respect to the history of science. For example, as Lee Bailey has so well demonstrated, not only was the camera obscura a favorite optical device in early Modern science, but it played a deliberately modeling role in Descartes's notion of both eye and ego (see Bailey 1989). From the camera obscura and its variants to the genuine photograph, there is a three-century history. This history finally focused upon the *fixing* of an image. As early as 1727, a German physician, Johann Schultze, did succeed in getting images onto chalk and silver powders, but the first successfully fixed image was developed by Joseph Niepce in 1826. His successor, Louis Daguerre, is credited usually (in 1839), but Daguerre simply perfected Niepce's earlier process (see Darius 1984, 34–35; Pollack 1977, 65–67). I shall jump immediately into the early scientific use of photography.

If the dramatic appearance of relative distance (space) was the forefront fascination with Galileo's telescope, one might by contrast note that it is the dramatic appearance of a transformation of *time* which photography brought to scientific attention. The photograph "stops time," and the technological trajectory implicitly suggested within it is the ever more precise micro-instant which can be captured. In early popular attention, the association with time stoppage often took the association between the depiction and a kind of "death" which still photography evoked. Ironically, the stilted and posed earliest photos were necessary artifacts of the state of the technology—a portrait could be obtained only with a minutes-long fixed pose, since it took that long for the light to form the negative on glass covered with the requisite chemical mixture.

Photography, however, was an immediately popular and rapidly developing new medium. And, if portraits and landscapes were early favored, a fascination with motion also occurred almost immediately. The pioneers of stop-motion photography were Eadweard Muybridge and Thomas Eakins at the end of the nineteenth century. Muybridge's studies of horses' gaits served a popular scientific interest. He showed, with both galloping horses and trotters, that all four feet left the ground, thus providing "scientific" evidence for an argument about this issue, considered settled with Muybridge's photos of 1878 (Darius 1984, 34–35). Insofar as this is a "new" fact (this is apparently debatable since there are some paintings which purport to show the same phenomenon), it is a discovery which is instrumentally mediated in a way parallel to Galileo's telescopic capture of mountains on their Moon. And, if this time-stop capacity of the technology can capture a horse's gait, the trajectory of even faster time-stop photography follows quickly. By 1888 time-stop photography had improved to the extent that the Mach brothers produced the first evidence of shock waves by photographing a speeding bullet. In this case, the photo showed that the bullet itself penetrated its target, not "compressed air," which was until then believed to advance before the projectile and cause injury (Darius 1984, 42–43).

Here we have illustrations of an early *perceptual hermeneutic* process which yields visually clear, repeatable, convincing Gestalts of the phenomena described. At this level, however, there is a "realism" of visual result which retains, albeit in a time-altered form, a kind of visual *isomorphism* which is a variant upon ordinary perception. It is thus less "textlike" than many other variants which develop later.

The visual isomorphism of early still photography was also limited to surface phenomena, although with a sense of frozen "realism" which shocked the artists and even transformed their own practices.[9] The physiognomy of faces and things was precise and detailed. The stoppage of time produced a *repeatable image of a thing*, which could be analytically observed and returned to time and again.

A second trajectory, however, was opened by the invention of the X-ray process in 1896. Here the "insides" of things could be depicted. Surfaces became transparent or disappeared altogether, and what had been "invisible" or, better, occluded became open to vision. X-ray photos were not so novel as to be the first interior depictions; we have already noted the invention of the "exploded diagram" style practiced by da Vinci and Vesalius. And one could also note that various indigenous art, such as that of Arnhemland Aborigines and Inuit, had an "X-ray" style of drawing which sometimes showed the interiors of animals. But the X-ray photo did to its objects what still photography had done to surfaces—it introduced a time-stop, "realistic" depiction of interior features. In this case, however, the X-ray image not only depicts differently, but produces its images as a "shadow." The X-rays pass through the object, with some stopped by or reduced by resistant material—in early body X-rays, primarily bones (see Kevles 1997, 3–20).

Moving rapidly, once again a trajectory may be noted, one which followed ever more distinct depiction in the development of the imaging technologies: today's MRI scan, CT tomography, PET scans, and sonograms all are variants upon the depiction of interiorities. Each of these processes not only does its depicting by different means but also produces different visual selectivities which vary what is more or less transparent and what is more or less opaque. (I shall return to these processes in more detail.) This continues to illustrate the inscription or visualization process which constitutes the perceptual hermeneutic style of science.

A third trajectory in visualization is one which continues from the earliest days of optical instrumentation: the movement to the ever more microscopic (and macroscopic) entities. The microscope was much later to find its usefulness within science than the telescope. As Ian Hacking has pointed out, as late as 1800 Xavier Bichet refused to allow a microscope in his lab, arguing that "When people observe in conditions of obscurity each sees in his own way and according as he is affected" (Hacking 1983, 193). In part, this had to do with the features of the things to be observed. Many micro-organisms were translucent or transparent and hard to make stand out even as figures against the often fluid grounds within which they moved. When another device which "prepared" the object for science was invented, *staining processes through aniline dyes*, the microscope could be more scientifically employed (Hacking 1983, 193; see ch. on "Microscopes," 186–209).

The trajectory into the microscopic, of course, explodes in the nineteenth and twentieth centuries, with electron microscopes, scanning, tunneling processes, and onto the processes which even produce images of atoms and atom surface structure. Let me include here, too, the famous radio chrystallography which brought us DNA structure, as well as today's chromosome and genetic fingerprinting processes.

The counterpart, macro-imaging, occurs with astronomy and the "earth sciences" which develop the measuring processes noted in chapter 4 [of *Expanding Hermeneutics*], concerning "whole Earth measurements." While each trajectory follows a different, exploitable image strategy, the result retains the Gestalt-charactered visualization which is a favored perceptual object within science.

The examples noted above all retain repeatable Gestalt, visualizable, and in various degrees, isomorphic, features. This is a specialized mode of perception and perceptual hermeneutics which plays an important role within science, but which also locates this set of practices within a now complicated lifeworld.

"Textlike" Visualizations

I shall now turn to a related, but different, set of visualizations, visualizations which bear much stronger relations to what can be taken as "textlike" features. Again, Latour is relevant: if the laboratory is science's *scriptorium*, the place where inscriptions are produced, then some of the production is distinctly textlike. A standard text, of course, is perceived. But to understand it one must call upon a specific hermeneutic practice—*reading*, and the skills which go in to reading.

Once again, it would be tempting to follow out in more detail some of the history of the writings which have made up our civilizational histories (and which characterize the postmodern penchant for textuality, following Derrida's *On Grammatology*). But as far as written texts are concerned, I want to note in passing only that the histories of writing have tended to converge into an ever narrowing set of choices: alphabetical, ideographic, and, for special purposes, simple pictographic forms. Related to this shrinkage of historical forms, I also want simply to note in passing that science follows and exacerbates this trend within its own institutional form, so much so that its dominantly alphabetic actual text preference is even more clearly narrowed to the emerging dominance of English as "the" scientific natural language.

And there is plenty of "text" in this sense within science. The proliferation of journals, electronic publications, books, and the range of texts produced is obvious enough. These texts, however, always remain secondary or tertiary with respect to science, as we have seen from Latour. So this is not the textlike phenomenon I have in mind; instead, I am pointing to those analogues of texts which permeate science: charts, graphs, models, and the whole range of "readable" inscriptions which remain visual, but which are no longer isomorphic with the referent objects or "things themselves."

Were we to arrange the textlike inscriptions along a continuum, from the closest analogue to the farthest and the most abstractly disanalogous, one would find some vague replication of the history of writing. Historians of alphabetic writing, for example, have often traced the letters of alphabetic writing to earlier *pictographic* items in pre-alphabetic inscriptions. Oscar Ogg, for example, shows that our current letter "A" derives from an inverted pictograph of a bull image (Ogg 1967, 78). Earlier hieroglyphic inscriptions could serve double purposes: as an analog image of the depicted animal or as the representation of a particular phoneme in the alphabetic sense.

The vestigial analog quality noted in the history of writing also occurs in scientific graphics: for example, a typical "translation" technology occurs in oscillography. If a voice is being patterned on an oscilloscope, the sound is "translated" into a moving, squiggly line on the scope. Each sound produces a recognizable squiggle, which highly skilled technicians can often actually "read."[10] The squiggle is no more, nor no less, "like" the sound made than the letter is within a text, but the technical "hermeneutic" can read back to the referent. As the abstraction progresses, often purposefully so that a higher degree of *graphic Gestalt* can be visualized, the reading-perception becomes highly efficient. "Spikes" on a graph, anomalies, upward or downward scatters—all have immediate significance to the "reader" of this scientific "text." Here is a hermeneutic process within normal science. And it remains visualizable and carries now in a more textlike context the repeatable Gestalt qualities noted above as part of the lingua franca of this style of hermeneutic.

Older instrumentation often was straightforward analogous to the phenomenon being measured. For example, columns of mercury within a thermometer embodied the "higher" and "lower" temperatures shown. Or, if a container was enclosed, a glass tube on the outside with piping to the inside could show the amount of liquid therein. Even moves to digital or numeric dials often followed analog representations.

Finally, although I am not attempting comprehensiveness in this location of hermeneutics in my "weak program" within scientific practice,

I want to conclude with some *conventions* which also serve to enhance the textlike reading perceptions. Graphs come with conventions: up and down for high and low temperatures or intensities; with the range of the growing uses of "false color" imagery, rainbow spectrum conventions are followed again for intensities, and so on. All of this functions "like" a reading process, a visual hermeneutics which retains its visualizations, but which takes textlike directions.

Summary of the "Weak Program"

In the weak program I have been following to this point, I have chosen science activities which clearly display their hermeneutic features. These, I have asserted, include a preference for visualization as the chosen sensory mode for getting to the things. But, rather than serving simply as a reduction of perceptual richness by way of a monosensory abstraction, visualization has been developed in a hermeneutic fashion—akin to "writing" insofar as writing is also a visual display. Thus, if science is separate from the lifeworld, it is so in precisely the same way that writing would not be included as a lifeworld factor.

Second, I have held that the process within science practice which prepares things for visualization includes the instrumentarium, the array of technologies which can produce the display, depiction, graphing, or other visualizable result which brings the scientific object "into view." (I am not arguing, as some have, that *only* instrumentally prepared objects may be considered to be scientific objects. But, in the complex late Modern sciences, instrumentation is virtually omnipresent and dominant when compared to the older sciences and their observational practices.)

Third, I am not arguing that these clearly hermeneutic practices within science *exhaust* the notion of science. I have not dealt with the role of mathematization, with forms of intervention which do not always yield visualizable results, or the need to take apart the objects of science, to analyze things. And I do not mean to imply that these factors are also important to science. Rather, I have been making, so far, the weak case that there are important hermeneutic dimensions to science, especially relevant to the final production of *scientific knowledge.* In short, hermeneutics occurs inside, within, science itself.

Moving now from the implicit hermeneutics within science praxis to the more complex practices—increasingly technologically embodied and instrumentally constructed—we are ready to take note of a stronger program.

The "Strong Program": Hermeneutic Sophistication

In the "weak program" I chose to outline what could easily be recognized as hermeneutic features operative within science. As I now turn to a stronger program, I shall continue to examine certain extant features within science practice which relate to hermeneutic activity, but I shall increasingly turn here to forefront modes of investigation which drive the sciences closer to a postmodern variant upon hermeneutics.

Whole Body Perception

It is, however, also time to introduce more fully, albeit sketchily, a phenomenological understanding of perception in action. This approach will be recognizably close to the theory of perception developed by Merleau-Ponty, although taken in directions which include stronger aspects of multistability and polymorphy, which earlier investigations of my own developed.

1. I have already noted some perceptual *Gestalt features*, including the presentation of a perceptual field, within which figure/ground phenomena may be elicited. Following a largely Merleau-Pontean approach, one notes that fields are always complexly structured, open to a wide variety of intentional interests, and bounded by a horizonal limit. Science, I have claimed, in its particular style of knowledge construction, has developed a visualist hermeneutic which in the contemporary sense has fulfilled its interests through *imagery* constituted instrumentally or technologically. The role of repeatable, Gestalt patterns, in both isomorphic and graphic directions, is the epistemological product of this part of the quest for knowledge.

2. In a strong sense, all sensory fields, whether focused upon in reduced "monosensory" fashion, or as ordinarily presented in synthesized and multidimensional fashion, are *perspectival* and concretely *spatial-temporal*. Reflexively, the embodied "here" of the observer not only may be noted but is a constant in all sensory perspectivalism. This constant may be enhanced only by producing a string of interrelated perspectives; or

by shifting into multiperspectival modes of observation. The "ideal observer," a "god's eye view," and nonperspectivalism do not enter a phenomenology of perception.

3. However, while a body perspective relative to the perceptual field or "world" is a constant, both the field and the body are *polymorphic* and *multistable*. In my work in this area, I have shown that multistability is a feature of virtually every perceptual configuration (and the same applies to the extensions and transformations of perception through instrumentation), and that the interrelation of bodily (microperceptual features) and cultural significations (macroperceptual features) makes the polymorphy even more complex. *There is no perception without embodiment; but all embodiment is culturally and praxically situated and saturated.*

4. While I have sometimes emphasized spatial transformations (Galileo's telescope) in contrast to temporal transformations (still photography), all perceptual spatiality is *spatial-temporal*. This space-time configuration may be shown with different effects, as in contrasts between visual repeatability and auditory patterning, but is a constant of all perceptions.

5. All perceptual phenomena are synesthetic and multidimensioned. The "monosensory" is an abstraction—although useful and possible to forefront—and simply does not occur in the experience of the "lived body" (*corps vécu*). The same applies, although not always noted, in our science examples. I will say more below on this feature of perception. The issue of the "monosensory" is particularly acute with respect to the technological embodiments of science, since instruments (not bodies) may be "monosensory." Again, we reach a contemporary impasse which has been overcome only in part. Either we turn ingenious in the ways of "translating" the spectrum of perceptual phenomena into a visual hermeneutic—perhaps the dominant current form of knowledge construction in science—or we find ways of enhancing our instrumental reductions through variant instruments or new modes of perceptual transformation (I am pointing to "virtual reality" developments here).

A "strong program," I am hinting, may entail the need for breakthroughs whereby a fuller sense of human embodiment may be brought into play in scientific investigations. Whereas the current, largely visualist hermeneutic within science may be the most sophisticated such mode of knowledge construction to date, it remains short of its full potential were "whole body" knowledge made equally possible. This would be a second step toward the incorporation of lifeworld structures within science praxis.

Instrumental Phenomenological Variations

In the voice metaphor I used to describe the investigation of things, I noted that the "giving of a voice" entails, actually, the production of a "duet" at the least. But this also means that different soundings may be produced, either in sequence or in array, by the applications of different instruments. This is a material process which incorporates the practice of "phenomenological variations" along with the intervention within which a thing is given a voice. This practice is an increasing part of science practice and is apparent in the emergence of a suite of new disciplines which today produce an ever more rapid set of revolutions in understanding or of more frequent "paradigm shifts."

I use this terminology because it is a theme which regularly occurs in science reporting. A Kuhnian frame is often cast over the virtually weekly breakthroughs which are reported in *Science, Scientific American, Nature*, and other magazines. Challenges to the "standard view" are common. I shall look at a small sample of these while relating the challenges to the instrumental embodiments which bring about the "facts" of the challenges. Here the focus is upon *multiple instrumental arrays* which have different parameters in current science investigation.

1. *Multiple new instruments/more new things.* The development of mutivariant instruments has often led to increased peopling of the discipline's objects. And much of this explosion of scientific ontology has occurred since the mid-twentieth century. This is so dramatically the case that one could draw a timeline just after World War II, around 1950 for most instrumentation, and determine new forms for many science disciplines. For example, in astronomy, until this century, the dominant investigative instrumentation was limited to *optical* technologies and thus restricted to the things which produce *light*. With the development of *radio-telescopy*, based upon technologies developed in World War II—as so many fields besides astronomy also experienced—the field expanded to the forms of microradiation which

occur along spectra beyond the bounds of visible light. *The New Astronomy* makes this "revolution" obvious. The editors note that

> The range of light is surprisingly limited. It includes only radiation with wavelengths 30 per cent shorter to 30 per cent longer than the wavelength to which our eyes are most sensitive. The new astronomy covers radiation from extremes which have wavelengths less and one thousand-millionth as long, in the case of the shortest gamma rays, to over a hundred million times longer for the longest radio waves. To make an analogy with sound, traditional astronomy was an effort to understand the symphony of the Universe with ears which could hear only middle C and the two notes immediately adjacent. (Henbest and Marten 1996, 6)

Without noting which instruments came first, second, and so on, expanding out from visible light, first to ultraviolet on one side, and infrared on the other, now reaching into the previously invisible-to-eyeball perceptions, but still within the spectrum of optical light waves, the first expansion in to invisible light range occurs through types of "translation" technologies as I have called them. The usual tactic here is to "constitute" into a *visible depiction* the invisible light by using some convention of *false color* depiction.

The same tactic, of course, is used once the light spectrum itself is exceeded. While some discoveries in radio astronomy were made by *listening* to the radio "hiss" of background radiation, it was not long before the gamma-to-radio wavelengths beyond optical capacities were also "translated" into visible displays.

With this new instrumentation, the heavens begin to show phenomena previously unknown but which are familiar today: highly active magnetic gas clouds, radio sources still invisible, star births, supernovas, newly discovered superplanets, evidence of black holes, and the like. The new astronomy takes us closer and closer to the "birth" of the universe. More instruments produce more phenomena, more "things" within the universe.

The same trajectory can be found in many other science disciplines, but for brevity's sake I shall leave this particular example as sufficient here.

2. *Many instruments/the same thing.* Another variant, now virtually standard in usage, is to apply a range of instruments which measure different processes by different means to the same object. Medical imaging is a

good example here. If some feature of the brain is to be investigated, perhaps to try to determine without surgical intrusion whether a formation is malignant or not, multiple instrumentation is now available to enhance the interpretation of the phenonomenon. A recent history of medical imaging, *Naked to the Bone*, traces the imaging technologies from the inception of X-rays (1896) to the present.

As noted above, X-rays allowed the first technologized making of the body into a "transparent" object. It followed the pattern noted of preparing the phenomenon for a scientific "reading" or perception. Early development entailed—to today's retrospective horror—long exposures, sometimes over an hour, to get barely "readable" images sometimes called "shadowgraphs." This is because X-ray imaging relies upon radiation sent through the object to a plate, and thus the degrees of material resistance cast "shadows" which form the "picture." The earliest problems focused upon getting clearer and clearer images (Kevles 1997, 20).

I have noted that the microscope became useful only when the specimen could be prepared for "reading" through a dye process which enhanced contrasting or differentiated structures (in the micro-organism). This image enhancement began to occur in conjunction with X-rays as early as 1911 with the use of *radioactive tracers* which were ingested or injected into the patient. This was the beginning of nuclear medicine (Kevles 1997, 304).[11]

Paralleling X-ray technologies, *ultrasound* began to be explored with the first brain images produced in 1937 (310). The quality of this imaging, however, remained poor since bone tended to reduce what could be "seen" through this sounding probe. (As with all technologies, it takes some time before the range of usefulness is discovered appropriate to the medium. In the case of ultrasound, soft tissue is a better and easier-to-define target.)

But, even later than the new astronomy, the new medical imaging does not actually proliferate into its present mode until the 1970s. Then, in 1971 and 1972, several patents and patent attempts are made for magnetic resonance processes (MRIs) (314). These processes produce imagery by measuring molecular resonances within the body itself. At the same moment, the first use of the computer—*as a hermeneutic instrument*—comes into play with the refinement of *computer-assisted tomography* (CT scanning). Here highly focused X-ray beams are sent through the object (brains, at first), and the data are stored and reconstructed through computer calculations and processes. Computer-"constructed" imaging, of course, began in the space program with the need to turn data into depictions. Kevles notes that "After the Apollo missions sent back computer-reconstructed pictures of the moon, it did not stretch the imagination to propose that computers could reconstruct the images of the interior of the body, which, like pictures from space, could be manipulated in terms of color and displayed on a personal video moniter" (143–44). Here mathematics and imagery or constructed perceivable depictions meet. I claim this is important to a strong hermeneutic program in understanding science.

By 1975 the practical use of positron emissions is captured in the PET scan process. These emissions (from positrons within the object) are made visible (314). This imagery has never attained the detail and clarity of the above technologies but has some advantage in a dynamic situation when compared to the "stills" which are produced by all but ultrasound processes (and which also are limited in clarity). Thus living brain *functions* can be seen through PET instrumentation. Then, in the 1990s, functional MRI and more sophisticated computer tomographic processes place us into the rotatable, three-dimensional depictions which can be "built up" or "deconstructed" at command, and the era of the *whole body image* is attained (314).

While each of these processes can show different phenomena, the multiple use is such that ever more complete analysis can also be made of single objects, such as tumors, which can be "seen" with differences indicating malignancy or benignity. This, again, illustrates the ever more complex ways in which science instrumentation produces a visible result, a visual hermeneutics which is the "script" of its interpretive activity.

3. *Many instruments/convergent confirmations.* Another variant upon the multiple instrument technique is to use a multiplicity of processes to check—for example, dating—for greater agreement. In a recent dating of Java *homo erectus* skulls, uranium series dating of teeth, carbon 14, and electron spin resonance techniques were all used to establish dates much more recent (27,000 B.P. +/– 53,000 B.P. for different skulls) than previously determined and thus found *homo erectus* to be co-extant with *homo sapiens sapiens* (*Science*, December 13, 1996, 1870–73). And, with the recent discovery of 400,000 B.P. javelin-like spears in Germany, one adds thermo-luminescence techniques to establish this new date for human habitation in Europe, at least double or triple the previously suspected earliest date for humans there. (Similar finds, now dated 350,000 B.P. in Siberia, and 300,000+ B.P. finds in Spain, all within 1996–97 discovery parameters, evidence this antiquity [*Science*, May 30, 1997, 1331–34].)

4. *Single instrument (or instrumental technique)/widespread multiple results.* Here perhaps one of the most widely used new techniques involves DNA "finger printing," which is now used in everything from forensics

(rapists and murderers both convicted and found innocent and released), to pushing dates back for human migrations or origins. (The "reading" process which goes with DNA identification entails matching pairs and includes visualizations once again.) *Scientific American* has recently reported that DNA tracing now shows that human migrations to the Americas may go back to 34,000 years (not far from the dates claimed for one South American site, claimed to be 38,000 B.P., which with respect to physical data remain doubtful), with other waves at 15,000 B.P., to more recent waves, included a set of Pacific originated populace around 6,500 B.P.) (*Scientific American*, August 1997, 46–47). The now widely cited DNA claims for the origins of *homo sapiens* between 200,000 and 100,000 B.P. is virtually a commonplace.

DNA fingerprinting has also been used in the various biological sciences to establish parentage compared to behavioral mating practices. One result is to have discovered that many previously believed-to-be "monogamous" species are, in fact, not. Similarly, the "Alpha Male" presumed successful at conveying his genes within territorial species has been shown to be less dominantly the case than previously believed (*Science*, March 31, 1989, 1663).

5. *Multiple instruments/new disciplines.* Beginning with DNA (mito-chondrial DNA matching) again, the application of this technique has given rise to what is today called "ancient DNA" studies, with one recent result developed in Germany this year, which purports to show that Neanderthals could not have interbred with modern humans, due to the different genetic makeup of these hominids who coexisted (for a time) with modern humans (*Science*, July 11, 1997, 176–78).

Then, returning to variants upon imaging, the new resources for such disciplines as archeology produce much more thorough "picturing" of ancient sites, activities, and relations to changes in environmental factors. Again, drawing from *Scientific American*—one can draw similar examples from vir-tually every issue of this and similar science-oriented magazines—the array of instruments now available produces, literally, an "in-depth" depiction of the human past. In part, now drawing from uses originally developed for military purposes, imaging from (1) *Landsat*, which used digital imaging and multispectral scanners from the 1970s, to (2) refinements for *Landsats 4* and *5* in the 1980s, which extended and refined the imaging and expanded to infrared and thermal scanning, (3) to *SPOT*, which added linear-array technologies to further refine imaging, to (4) imaging radar, which actually penetrates below surfaces to reveal details, to (5) *Corona*, which provides spectrographic imagery (recently declassified), the modern archeologist, particularly desert archeologists, can get full-array depictions of lost cities, ancient roads, walls, and the like from remote sensing used now (*Scientific American*, August 1997, 61–65).

Once located, instrumentation on Earth includes (1) electromagnetic sounding equipment, which penetrates up to six meters into the earth, (2) ground-penetrating radar, which goes down to ten meters, magnetometers, which can detect such artifacts as hearths, (4) resistivity instruments, which detect different densities and thus may be used to locate artifacts, and (5) seismic instruments, which can penetrate deeper than any of the above instruments. At a recent meeting in Mexico, an anthropologist reported to me that a magnetometer survey of northern Mexico has shown there to be possibly as many as 86,000 buried pyramids (similar to the largest, Cholula, although the remainder are smaller).[12]

In short, the proliferation of instrumentation, particularly that which yields imagery, is radical and contemporary and can now yield degrees and spans of three-dimensional imagery which includes all three of the image breakthroughs previously noted: early optics magnified the micro- and macro-aspects of barely noted or totally unnoted phenomena through magnification (telescopy and microscopy) but remained bound to the limits of the optically visible, by producing "up close" previously distant phenomena.

Photography increased the detail and isomorphy of imaging in a repeatable produced image, which could then be studied more intensely since "fixed" for observation. It could also be manipulated by "blowups" and other techniques, to show features which needed enhancement. Then, with X-rays, followed by other interior-producing imagery, the possibilities were outlined for the contemporary arrays by which image surfaces are made transparent so that one may see interiors. Then add the instruments which expand thoroughly beyond the previously visible and which now go into previously invisible phenomena through the various spectra which are "translated" into visible images for human observation.

Here we have the decisive difference between ancient science and Modern science: Democritus claimed that phenomena, such as atoms, not only were in fact imperceptible, but were *in principle* imperceptible. A modern, technologized science returns to Democritus to the "in fact" only—that is, the atomic is invisible only until we can come up with the technology which can make it visible. This, I claim, is an instrumental *visual hermeneutics*.

Technoconstitution

In the reconstrual of science which I have been following here, I have argued that late Modern science has developed a complex and sophisticated system of *visual hermeneutics*. Within that visualist system, its "proofs" are focused around the things seen. But, also, things are never just or merely seen—the things are *prepared* or made "readable." Scientifically, things are (typically, but not exclusively) *instrumentally mediated,* and the "proof" is often a *depiction* or image.

Interestingly, if in ordinary experience there is a level of *naive realism* where things are taken simply to be what they are seen to be, similarly, within imaging there is at least the temptation to an imaging naive realism. That naïveté revolves around the intuitive taking of the image to be "like" or to "represent" an original (which would be seen in unmediated and eyeball perceptions). In short, "truth" is taken to be some kind of *isomorphism* between the depiction and the object.

By putting the issue in precisely that way, there are many traps which are set which could lead us back into the issues of modern (that is, Cartesian or seventeenth- and eighteenth-century) epistemologies. But to tackle these would lead us into a detour of some length. It would entail deconstructing "copy theory" from Plato on, deconstructing "representationalism" as the modern version of copy theory, before finally arriving at a more "postmodern" theory which entails both a theory of relativistic intentionality, a notion of perspectivalism, and an understanding of instrumental mediations as they operate within a phenomenological context. (I have addressed these issues in some degree in essays which preceded my formulation here.)[13] I simply want to avoid these traps.

To do so, I shall continue to interpret science in terms of a visual hermeneutics, embodied within an instrumentally realistic—but *critical*—framework in which instruments mediate perceptions. The device I shall now develop will fall within an idealized "history" of imaging, which, while containing actual chronologically recognizable features, emphasizes patterns of *learning to see*.

Isomorphic Visions

The first pattern is one which falls into one type of *initial isomorphism* within imaging. As a technical problem, it is the problem of getting to a "clear and distinct" image. Imaging technologies do not just happen, they develop. And in the development there is a dialectic between the instrument and the user in which both a learning-to-see meets an elimination-of-bugs in the technical development. This pattern is one which, in most abstract and general terms, moves from initial "fuzziness" and ambiguity to greater degrees of clarity and distinctness.

Histories of the telescope, the microscope, photography, and X-rays (and, by extension, all the other imaging processes as well) are well documented with respect to this learning-to-see. Galileo, our quasi-mythical founder of early Modern science, was well aware of the need to teach telescopic vision, and of the problems which existed—although he eventually proclaimed the *superiority* of instrumentally mediated vision over ordinary vision. The church fathers, however, did have a point about how to take what was seen through the telescope. Not all of Galileo's observations were clear and easily seen by "any man." The same problem reemerged in the nineteenth century through the observations of Giovanni Schiaparelli, who gave the term the "canals of Mars." Schiaparelli was a well-known astronomer who had made a number of important discoveries, particularly with respect to asteroids and meteor swarms (because, in part, he had a much better telescope than Galileo). But in noting "canali"—which should have been translated into "channels"—which were taken to be "canals"—he helped stimulate the speculations about life on Mars. But neither channels nor canals existed—these, too, were instrumental artifacts (*Micropaedeia, Encyclopaedia Britannica*, 15th ed., 10:514).

The dialectic between learning and technical refinement, in the successful cases, eventually leads to the production of clear and distinct images and to quick and easy learning. These twin attainments, however, cover over and often occlude the history and struggle which preceded the final plateaus of relative perfections. Thus, as in the previous illustrations concerning my guests and our Vermont observations of the Moon, once focused and set, it literally takes only instants before one can recognize nameable features of its surface. The "aha phenomenon," in short, is virtually immediate today because it is made possible by the advanced technologies. That instantaneity is an accreted result of the hidden history of learning-to-see and its accompanying technical debugging process.

This same pattern occurred with the microscope. Although microorganisms never before seen were detected early, the continued problems of attaining clear and distinct microscopic vision was so difficult that it did not allow the microscope to be accepted into ordinary scientific practice until the nineteenth century. Again, the dialectic of learning how and what to see meets the gradual technical improvement concerning lenses and focusing devices, and finally the application of dying procedures to the things themselves. (This is an overt example of preparing a thing to become a scientific or "readable" object!)

Photography stands in interesting contrast to microscopy—if it took a couple of centuries for microscopy to become accepted for scientifically acceptable depiction, photography was much faster to win the same position. From Niepce's first "fixed" image in 1826, to the more widely accepted date of 1839 for Daguerre's first images, it was less than a half century until, as Bettyann Kevles notes in her history of medical imaging, "By the 1890's photographs had become the standard recorders of objective scientific truth" (Kevles 1997, 15).

The same pattern occurs, but with even greater speed, in the history of early X-rays. In publicizing his new invention, Wilhelm Röntgen made copies of the X-ray of his wife's hand, which showed the bones of her fingers and the large ring which she wore, and sent these to his colleagues across Europe as evidence of his new process. That X-ray (with a long exposure time) was fuzzy, and while easily recognizable as a skeletal hand and ring, contrasts starkly with the radiograph made by Michael Pupin of Prescott Butler's shot-filled (shotgun-injured) hand later the same year (21, 35). X-rays, duplicated across Europe and America almost immediately after Röntgen's invention, were used "scientifically" from the beginning.

The acceleration of acceptance time (of the learning-technical vision dialectic) similarly applies to the recent histories of imaging, which include, as above, sonograms (1937) and MRIs (1971) in medicine, of remote imaging since the *Tiros* satellite (1965), or of digitally transmitted and reconstructed images from *Mariner 4* (1965) in Earth and space science.

All of the above samples, however, remain within the range of the possible "naive image realism" of visual isomorphisms in which the objects are easily recognizable, even when new to the observer's vision. (Even if Röntgen had never before seen a "transparent hand" as in the case of his wife's ringed fingers, it was "obvious" from the first glimpse what was seen.) The pattern of making clear is an obvious trajectory. Yet we are not quite ready to leave the realm of the isomorphic.

How does one make "clearer" what is initially "fuzzy"? The answer lies in forms of manipulation, what I shall call *image reconstruction*. The techniques are multiple: enlargements (through trajectories of magnification noted before), enhancements (where one focuses in upon particular features and finds ways to make these stand out), contrasts (by heightening or lessening features of or around the objects), and so on. In my examinations I shall try not to be comprehensive, but to remain within the ranges of familiarity (to at least the educated amateur) concerning contemporary imaging. All of these manipulations can and do occur within and associated with simply isomorphic imaging and, for that matter, within its earlier range of black-and-white coloring. Histories of the technical developments which go with each of these techniques are available today and provide fascinating background to the rise of scientific visualism.

The moral of the story is images don't just occur. They are made. But, once made—assuming the requisite clarity and accuracy and certification of origin, etc.—they may then be taken as "proofs" within the visual hermeneutics of a scientific "visual reading." We are, in a sense, still within a Latourean laboratory.

Translation Techniques

Much of what can follow in this next step has already been suggested within the realm of the isomorphic. But what I want to point to here is the use in late Modern science of visual techniques which begin ever more radically to vary *away from the isomorphic.*

One of these variables is—if it could be called that—simply the variable use of *color.* Returning to early optics, whatever Galileo or Leeuwenhoek saw, they saw in "true color." And, as we have seen, sometimes that itself was a problem. The transparent and translucent micro-organisms in "true color" were difficult to see. With aniline dyes, we have an early use of "false color." To make the thing into a scientific or "readable" object, we intervene and create a "horse of a different color." "False coloring" becomes a standard technique within scientific visual hermeneutics.

The move away from isomorphism, taken here in gradual steps which do not necessarily match chronologically what happened in the history of science, may also move away from the limits of ordinary perception. As noted above, the "new" late Modern astronomy of midcentury to the present was suddenly infused with a much wider stretch of celestial "reality" once it moved beyond optical and visible limits into, first, the humanly invisible ranges of the still optical or light itself, in the ranges of the infrared and ultraviolet. The instrumentation developed was what I have been calling a *translation technology* in that the patterns which are recordable on the instrumentation can be rendered by "false coloring" into visible images. This same technique was extended later to the full wave spectrum now available from gamma rays (short waves) through the optical to radio waves (long waves), which are rendered in the standard visually gestaltable, but false color, depictions in astronomy. All this is part of the highly technologized, instrumentalized visual hermeneutics which makes the larger range of celestial things into seeable scientific objects.

The "realism" here—and I hold that it is a realism—is a Hacking-style realism: if the things are "paintable"[14] (or "imagable") with respect to what the instruments detect as effects which will not go away, then they are "real." But they have been *made visible* precisely through the technological constructions which mediate them.

Higher Level Constructions

Within the limits of the strong program, I now want to take only two more steps: I am purposely going to limit this attempt to reconstrue science praxis as hermeneutic to contemporary imaging processes which make (natural) things into scientific, and thus "readable," visual objects. I am not going to address the related, but secondary, visual process which entails *modeling*. That process which utilizes the computer as a hermeneutic device is clearly of philosophical interest, but I shall stop short of entering that territory here.

Computers, of course, are integral to many of the imaging processes we have already mentioned. Medical *tomography* (MRI, PET, fMRI, etc.) entails computer capacities to store and construct images. What is a visual Gestalt is built up from linear processes which produce data which have to be "constructed" by the computer. Similarly, the digitally transmitted imagery from distance sensing in satellite, space, and other remote imaging processes also has necessary computer uses. Much of contemporary imaging is computer embodied. And computers open the ways to much more flexible, complex, and manipulable imaging than any previous technology. For the purposes here, however, they will remain simply part of the "black boxes" which produce images which mediate perceptions.

The two higher-level constructive activities I want to point to here entail, first, the refinement of imaging which can be attained through specifically recognizing our technologies as mediating technologies which, in turn, must take into account the "medium" through which they are imaging. I turn again to astronomical imaging: the *Hubble* space telescope has recently captured the most public attention, but it is but one of the instrumental variations which are today exploring the celestial realms.

The advantage *Hubble* has is that it is positioned beyond the effects of the atmosphere with its distortions and interferences—the clarity of *Hubble* vision in this sense is due in part to its extra-Earthly perspective. (Science buffs will recall that at launch it had several defects in operation which were subsequently fixed—thus placing the *Hubble* in the usual pattern of needing technical adjustment to make its images clear!) But, in part by now being able to (phenomenologically) vary *Hubble* with Earth-bound optical telescopes, the move to enhancing Earth-bound telescopy through computer compensations has become possible. Astronomy is moving toward technoconstructions which can account for atmospheric distortions "on the spot" through a combination of laser targeting and computer enhancements. Earth-bound telescopes are today being given new life through these hi-tech upgrades which "read" atmospheric distortions and "erase" these processes which can make clearer new "readable" images. *Science* regularly publishes an "imaging" issue devoted to updating what is taken as the state-of-the-art in imaging (in 1997 it was the 27 June issue). A description of how one "undoes the atmosphere" is included, which entails computer reconstructions, telescopes in tandem, and adaptive optics. This process, *Science* claims, "combat[s] the warping effects of gravity on their giant mirrors . . . reclaims images from the ravages of the atmosphere . . . [and] precisely undoes the atmospheric distortions" (*Science*, June 27, 1997, p. 1994).[15]

But alongside *Hubble* are the other variants: the infrared space observatory, the *Cosmic Background Explorer*, and other satellite instrumentation which produces imagery from the nonoptical sources. All these technologies are variants upon the same multidimensioned variables which produce readable images, or make things into readable scientific objects.

The final set of instrumental productions I wish to note are the *composites* which produce variants upon "wholes." Chapter 4 [of *Expanding Hermeneutics*] deals with "whole Earth measurements" which constitute one realm of composite imagery. To determine whether or not sea levels are rising overall, the composite imagery produced combines (1) multiple satellite photo imagery, (2) Earth-bound measurements (such as buoys, laser measurements, and land markers), and (3) computer averaging processes to produce a depiction (false colored) which can, in comparing time slices, show how much the oceans have risen. The composite depiction displays a flat-projection map of the Earth with level plateaus in false color spectra which can be compared between years, decades, and so on.

Similar processes occur in medical imaging. The "whole body imagery" available today on the internet is the result of two full-body "image autopsies," one each of a male and a female, whose bodies through tomographic processes may be seen in whatever "slice" one wishes. The linear processes of tomography show, slice-by-slice, vertically, horizontally, or in larger scans, the full bodies of the corpses used. The dimensions can be rotated, realigned, sectioned, and so on. Tomography also allows one to "peel," layer by layer, the object imaged—from skin, to networked blood vessels, to bones, and so on. (Both the whole Earth and whole body images are probably among the world's most expensive "pictures.") Moreover, all the manipulations which entail enhancements, contrasts, colorings, translations, and the like are utilized in these "virtual" images. Yet, while these virtual "realities" are different from the examination of any actual cadaver, they clearly belong to the visual hermeneutics of science in the strong sense. Things have been prepared to be seen, to be "read" within the complex set of instrumentally delivered visibilities of scientific imaging.

References

Bailey, Lee W. 1989. "Skull's Darkroom: The *Camera Obscura* and Subjectivity." In *Philosophy of Technology*, edited by Paul T. Durbin, 63–80. Dordrecht: Kluwer.
Brown, Harold I. 1985. "Galileo on the Telescope and the Eye." *Journal of the History of Ideas* 46:487–501.
Darius, Jon. 1984. *Beyond Vision*. Oxford: Oxford University Press.
Dreyfus, Hubert. 1993. *What Computers Still Can't Do: A Critique of Artificial Reason*. Cambridge, MA: MIT Press.
Galison, Peter. 1987. *How Experiments End*. Chicago: University of Chicago Press.
Hacking, Ian. 1983. *Representing and Intervening*. Cambridge: Cambridge University Press.
Henbest, Nigel, and Michael Marten. 1996. *The New Astronomy*. 2nd ed. Cambridge: Cambridge University Press.
Husserl, Edmund. 1970. *The Crisis of European Sciences and Transcendental Phenomenology*. Translated by David Carr. Evanston, IL: Northwestern University Press.
Ihde, Don. 1976. *Listening and Voice: A Phenomenology of Sound*. Athens: Ohio University Press.
———. 1986. *Experimental Phenomenology*. Albany: State University of New York Press.
———. 1990. *Technology and the Lifeworld: From Garden to Earth*. Bloomington: Indiana University Press.
———. 1993. *Postphenomenology: Essays in the Postmodern Context*. Evanston, IL: Northwestern University Press.
Kevles, Bettyann Holtzmann. 1997. *Naked to the Bone: Medical Imaging in the Twentieth Century*. New Brunswick, NJ: Rutgers University Press.
Kuhn, Thomas S. 1962. *The Structure of Scientific Revolutions*. Chicago: University of Chicago Press.

Merleau-Ponty, Maurice. 1962. *Phenomenology of Perception.* Translated by Colin Smith. London: Routledge and Kegan Paul.

Ogg, Oscar. 1967. *The 26 Letters.* New York: Crowell.

Pollack, Peter. 1977. *The Picture History of Photography.* New York: Abrams.

Van Sertima, Ivan. 1986. *Blacks in Science: Ancient and Modern.* New Brunswick, NJ: Transaction.

Notes

Introduction

1. To break it down further, the count of 25 monographs includes original books and collections of articles, as well as two expanded second editions in 2007's *Listening and Voice*, second edition, and 2012's *Experimental Phenomenology: Multistabilities*, second edition, as well as the book *Let Things Speak*, a 2008 volume in Chinese comprised of lectures from that time period. This also includes Ihde's series of what he calls his LLLBs (or "late life little books"), short article collections often with names in the form of "X Technics": for example, *Ironic Technics, Embodied Technics*, and *Medical Technics* (2008b, 2010a, 2019). In addition, this count includes the transcription of a 1986 seminar at the University of Gothenburg, published as short monograph entitled *On Non-foundational Phenomenology*, and one that is reproduced in its entirety here in this collection as chapter 3.

2. For examples of special issues of journals centered on Ihde's work, see Ihde 2008d; Selinger 2008; Goeminne and Paredis 2011; Rosenberger 2016; Ihde and Malafouris 2019; Botin, de Boer, and Børsen 2020; Lemmens and Van Den Eede 2022.

3. Ihde's deep engagement with the topic of imaging has laid the groundwork for many others to continue this research, both in terms of philosophical study and through empirical investigation, all deeply interdisciplinary and case study–oriented work. For an example of the continuing interest in these themes, see Fried and Rosenberger 2021.

Chapter 1

This chapter is from chapters 4 and 5 of *Listening and Voice: Phenomenologies of Sound*, by Don Ihde, 2nd ed. (Albany: State University of New York Press, 2007), 49–71.

1. A phenomenological warning must be issued here. There is a strict difference between empty supposing and what is intuitionally fulfilled. Thus the

exercise at this point is not strictly phenomenological but proceeds toward strict phenomenology by approximations.

2. Echo-location by clicks is more accurate than by tones. "If continuous tones were used instead of clicks there was a significant loss of accuracy in the perception of the direction of the sounds, though the experiences had the same character of wave lengths greater than 2K" (von Bekesy 1969, 287).

Chapter 2

This chapter is from chapters 4 and 5 of *Experimental Phenomenology: An Introduction*, by Don Ihde, 2nd printing (Putnam, 1977; Albany: State University of New York Press, 1986), 67–90.

Chapter 3

From *On Non-foundational Phenomenology*, by Don Ihde, ed. Seth Chaiklin, Fenomenografiska notiser 3 (Gothenburg: Institutionen för pedagogik, Göteborgs universitet, 1986).

The 1986 editor's preface by Seth Chaiklin states:

> The following text is an edited version of seminar given by Professor Don Ihde, State University of New York at Stony Brook, on 14 December 1984, at Institutionen för pedagogik, University of Gothenburg. I attempted to keep the informal quality of the spoken form, and have included questions from the audience as they occurred. These questions are identified as AX, where X identifies the questioner. [In this reedited version for *The Critical Ihde*, questions are now noted in italics and "Audience Questioner X," followed by Ihde's response.] I made some grammatical corrections and stylistic abbreviations and expansions. Substantive expansions are marked with parentheses. I added a title to the lecture, headings in the text, and references. Thanks to Jan Marton for the original transcription.

1. The original version of this piece also contains a short reflection on Vaserely cube paintings. These remarks have not been included here.

Chapter 4

From "Response to Rorty, Or, Is Phenomenology Edifying?," chapter 9 of *Consequences of Phenomenology*, by Don Ihde (Albany: State University of New York Press, 1986), 181–98.

1. Robert Nozick's *Philosophical Explanations* was the year's contender. But if my small sample is indicative, whereas Rorty was read, Nozick's tome rarely was finished by readers.

2. I continue the convention of the Introduction here with ACE meaning the American Continental Establishment and AE the Analytic Establishment.

3. Others are also aware of the need to recharacterize the practice of analytic philosophers, as in Moulton's use of a legal practice.

4. I have long contended that Husserl must be read-through. His heuristic discourses on method are attempts, after he has seen something difficult to see, to tell others how to do it. By adopting extant terminologies and then reversing or radically changing their meanings, his work is almost metaphorical. Heidegger, I would contend, must be read *literally.*

5. To term these gestalts and imply Husserl used them is a bit anachronistic since the Gestaltists were aware of and in some cases were students of Husserl.

6. Privately circulated, this list apparently came out of one of the Dreyfus summer programs.

7. A dissertation by Gary Aylesworth traces both the Wittgensteinian and Heideggerian directions carefully (1984).

8. See my "Phenomenology and the Later Heidegger," which shows the way in which phenomenology functions in his later works (in Ihde 1983, 119–36).

9. Chapter two of this collection [*Consequences of Pragmatism*] goes into more detail on Merleau-Ponty and Foucault. *Editor's Note:* More on Foucault can also be found here in *The Critical Ihde* in the previous chapter.

Chapter 5

From "What Is Postphenomenology?," chapter 1 of *Postphenomenology and Technoscience: The Peking University Lectures,* by Don Ihde (Albany: State University of New York Press, 2009), 5–23. There are several works mentioned in this chapter for which the author has not provided references.

Chapter 6

From "Program One: An Phenomenology of Technics," chapter 5 of *Technology and the Lifeworld,* by Don Ihde (Bloomington: Indianna University Press, 1990), 72–123.

1. This illustration is my version of a similar one developed by Patrick Heelan in his more totally hermeneuticized notion of perception in *Space Perception and the Philosophy of Science* (1983, 193).

2. Hacking (1983, 195) develops a very excellent and suggestive history of the use of microscopes. His focus, however, is upon the technical properties that were resolved before microscopes could be useful in the sciences. He and Heelan,

however, along with Robert Ackermann, have been among the pioneers dealing with perception and instrumentation in instruments. Cf. also my *Technics and Praxis* (Ihde 1979).

3. A more complete discussion of the specifics of Aboriginal ethics may be found in David H. Bennett's "Inter-Species Ethics: A Brief Aboriginal and Non-Aboriginal Comparison" (1985).

Chapter 7

From "Embodying Hearing Devices: Digitalization," chapter 22 of *Listening and Voice*, by Don Ihde, 2nd ed. (Albany: State University of New York Press, 2007), 243–50.

1. My own discussion of Heidegger's philosophy of technology, including the hammer analysis, may be found in *Technics and Praxis* (Ihde 1979, 103–29) and again in briefer form in *Technology and the Lifeworld* (Ihde 1990, 31–34).

2. The news item I refer to was from the *Independent*, read during a trip to the UK—I did not jot down the reference.

3. Steven Feld describes an acoustic-rich language among New Guinean highlanders in *Senses of Place* (Feld 1996).

4. An especially sensitive and phenomenologically insightful account of a high-tech limb may be found in Vivian Sobchack, *Carnal Thoughts* (2004).

5. See Bettyann Holzmann Kevles, *Naked to the Bone* (1997), also Finn Olesen on the stethoscope in *Postphenomenology: Critical Companion to Ihde* (2006).

Chapter 8

From "Deromanticizing Heidegger," chapter 8 of *Postphenomenology: Essays in the Postmodern Context*, by Don Ihde (Evanston, IL: Northwestern University Press, 1993), 103–15.

1. Heidegger (1982, 118–19). Translation by Michael Heim (1987).

Chapter 9

From "Technology and Cultural Variations," chapter 5 of *Consequences of Phenomenology*, by Don Ihde (Albany: State University of New York Press, 1986), 116–36.

Chapter 10

From "Technology and Prognostic Predicaments," by Don Ihde, in *AI and Society* 13 (1999): 44–51.

Chapter 11

From "Husserl's Galileo Needed a Telescope!," by Don Ihde, in *Philosophy and Technology* 24 (2011): 69–82.

1. "If '*positivism*' is tantamount to an absolutely unprejudiced grounding of all sciences on the 'positive,' that is to say, on what can be seized upon originaliter, then we are the genuine positivists. In fact, we allow no authority to curtail our right to accept all kinds of intuition as equally valuable legitimating sources of cognition—not even the authority of 'modern natural science' " (Husserl 1982, 39).

2. "In the intuitively given surrounding world . . . we experience 'bodies'—not geometrical-ideal bodies but precisely those bodies that we actually experience, with the content which is the actual content of experience" (Husserl 1970, 25).

3. I am taking as praxis the activities which constitute various meaning-regions, such as Husserl's "invention" of Egyptian surveying praxis: "The art of measuring discovers practically the possibility of picking out as [standard] measures certain empirical basic shapes, concretely fixed on empirical rigid bodies which are in fact generally available; and by means of these relations which obtain (or can be discovered) between these and other body-shapes it determines the latter intersubjectively and in practice univocally—at first within narrow spheres (as in the art of surveying land). . . . The art of measuring thus becomes the trail-blazer for the ultimately universal geometry and its 'world' of pure limit-shapes" (Husserl 1970, 28).

4. "The geometry which is ready-made, so to speak, from which the regressive inquiry begins, is a tradition. Our human existence moves within innumerable traditions. The whole cultural world, in all its forms, exists through tradition. . . . everything traditional has arisen out of human activity, that accordingly past men and human civilizations existed, and among them their first inventors, who shaped the new of out materials at hand, whether raw or already spiritually shaped" (Husserl 1970, 354–55. "For a genuine history of philosophy, a genuine history of the particular sciences, is nothing other than the tracing of the historical meaning structures given in the present, or their self-evidences, along the documented chain of historical back-references into the hidden dimension of the primal self-evidences which underlie them" (Husserl 1970, 372).

5. "Making geometry self-evident, then, whether one is clear about this or not, is the disclosure of its historical tradition" (Husserl 1970, 371). "The geometry of idealities was preceded by the practical art of surveying, which knew nothing of idealities. Yet such a pregeometrical achievement was a meaning-fundament for geometry . . ." (Husserl 1970, 49).

6. "Here is again something confusing: every practical world, every science, presupposes the life-world; as purposeful structures they are contrasted with the life-world, which was always and continues to be 'of its own accord.' Yet, on the other hand, everything developing and developed by mankind . . . is itself a piece of the life-world" (Husserl 1970, 382–83).

7. "But now we must note something of the highest importance that occurred even as early as Galileo: the surreptitious substitution of the mathematically substructed world of idealities for the only real world, the one that is actually given through perception, that is every experienced and experienceable—our everyday life-world" (Husserl 1970, 48–49).

8. The confusion of life-world and science as a piece of the life-world (see above) ". . . is only confusing because the scientists, like all who live communalized under a vocational end . . . have eyes for nothing but their ends and horizons of work. No matter how much the life-world is the world in which they live, to which even all their 'theoretical works' belong, and no matter how much they make use of elements of the life-world, which is precisely the 'foundation' of theoretical treatment as that which is treated, the lifeworld is just not their subject matter . . . and thus [their subject matter] is not, in the full survey, the universe of what is . . ." (Husserl 1970, 383).

9. ". . . The life-world is the world that is constantly pregiven, valid constantly and in advance as existing, but not valid because of some purpose of investigation, according to some universal end . . . scientific truth presupposes it . . . and in the course of [scientific] work it presupposes it every anew, as a world existing in its own way . . ." (Husserl 1970, 382).

10. "The scientific world . . . is a purposeful structure extending to infinity—a structure [made by] men who are presupposed, for the presupposed life-world" (Husserl 1970, 382).

11. "But the book cannot be understood unless one first learns to comprehend the language and to read the alphabet in which it is composed. It is written in the language of mathematics, and its characters are triangles, circles, and other geometric figures, without which it is humanly impossible to understand a single word of it; without these, one wanders about in a dark labyrinth" (Galileo, as quoted in Sobel 1999, 16).

12. "I render infinite thanks to God for being so kind as to make me alone the first observer of marvels kept hidden in obscurity for all previous centuries" (Galileo, as quoted in Sobel 1999, 6).

13. Father Clavius . . . laughing at Galileo's pretended four satellites of Jupiter, said he, too, could show them if he were only given time "first to build them into some glasses." Boorstin 1985, 316).

14. See a full discussion of this issue in Brown 1985.

15. Heidegger: "When we use the word 'science' today, it means something essentially different from the doctrina and scientia of the Middle Ages, and also from the Greek episteme. Greek science was never exact because in keeping with its essence, it could not be exact and did not need to be exact. Hence it makes no sense whatever to suppose that modern science is more exact than that of antiquity. Neither can we say that the Galilean doctrine of freely falling bodies is true or that Aristotle's teaching that light bodies strive upward is false; for the Greek

understanding of the essence of body and place and the relation between the two rests upon a different interpretation of beings and hence conditions a correspondingly different kind of seeing and questioning of natural events" (Heidegger 1997). Kuhn: "Since remote antiquity most people have seen one or another heavy body swinging back and forth on a string or a chain until it finally comes to rest. To the Aristoteleans, who believed that a heavy body is moved by its own nature from a higher position to a state of natural rest at a lower one, the swinging body was simply falling with difficulty. Galileo, on the other hand, looking at the swinging body, saw a pendulum, a body that almost succeeded in repeating the same motion over and over again ad infinitum. . . . I am acutely aware . . . of the difficulties created by saying that when Aristotle and Galileo looked at swinging stones, the first saw constrained fall, the second a pendulum. . . . when paradigms change, the world changes with the . . . paradigm changes . . . [which] cause scientists to see the world of their research engagement differently" (Kuhn 1962).

Chapter 12

From "The Historical-Ontological Priority of Technology over Science," chapter 2 of *Existential Technics*, by Don Ihde (Albany: State University of New York Press), 25–46.

1. Technology, capitalized indicates a use similar to what Heidegger calls the essence of *technology*.

2. "The current conception of technology, according to which it is a means and a human activity, can therefore be called the instrumental and anthropological definition of technology" (Heidegger 1977, 288).

3. "In a separate building outside Hagia Sophia, Justinian places a clepsydra and sundials, but clocks were never permitted within or on Eastern churches; to place them there would have contaminated eternity with time. As soon, however, as the mechanical clock was invented in the West, it quick spread not only to the towers of Latin churches but also to their interiors" (White 1972, 198).

4. A popular discussion of these techniques may be found in the 1974, 146, no. 6, issue of *National Geographic*.

5. I call these existential relations. See chapter one of my book *Technics and Praxis* (Ihde 1979).

Chapter 13

From "Art Precedes Science, or Did the *Camera Obscura* Invent Modern Science?," by Don Ihde, in *Instruments in Art and Science*, ed. Helmar Schramm, Ludger Schwarte, and Jan Lazardzig (Berlin: de Gruyter, 2008), 383–93.

1. Cf. Wilson 1995. In addition to being a history of the microscope, Wilson goes into detail about how it was taken by early modern philosophers, including Descartes and Locke. Regarding the familiarity of historically informed people concerning the *camera obscura* and its use by artists, in addition to a full article to this effect in the 1929 *Encyclopedia Britannica*, Peter Pollack's *The Picture History of Photography* (1977) has a brief but detailed and illustrated history of art and the camera—and of its development into the photographic version of the camera.

2. "Camera Obscura" in *Encyclopedia Britannica* (1929, 4:658).

Chapter 14

From "Scientific Visualism" and "Technoconstruction," chapters 12 and 13 of *Expanding Hermeneutics*, by Don Ihde (Evanston, IL: Northwestern University Press, 1998), 151–89.

1. Husserl distinguishes between the lifeworld and the "world" of science: "The life-world is the world that is constantly pregiven, valid constantly and in advance as existing. . . . [E]very science presupposes the life-world; as purposeful structures they are *contrasted* with the life-world, which was always and continues to be 'of its own accord'" (Husserl 1970, 382).

2. Merleau-Ponty continues the distinction made by Husserl: "The whole universe of science is built upon the world as directly experienced Science has not and never will have, by its nature, the same significance qua form of being as the world which we perceive. . . . [S]cientific points of view, according to which my existence is a moment of the world's, are always both naive and at the same time dishonest, because they take for granted, without explicitly mentioning it, the other point of view . . . through which from the outset a world forms itself round me and begins to exist for me" (Merleau-Ponty 1962, viii–ix).

3. This is a sustained thesis in my *Technology and the Lifeworld* (Ihde 1990).

4. Kuhn cites numerous examples of this problem, placing it in the context of changed perceptions—he claims that one Gestalt shift concerning "chaff particles" in the seventeenth century becomes "electrostatic repulsion" in the nineteenth, but he admits that "electrostatic repulsion was not seen as such until Hauksbee's large-scale apparatus had greatly magnified its effects" (Kuhn 1962, 117).

5. Galison argues that experiments do not end—rather, there are repeated endings and refinements which often result in a general agreement that phenomena are established when effects would not "go away" (see Galison 1987, 237).

6. My son, Mark, theorized that the claims were made by people who needed glasses. He noted that when he took his own glasses off, the Moon did appear smooth, so maybe the church fathers were the old men who could write—or, perhaps, the dark and light contrasts reflected Earth features?

7. Harding's reference is to Van Sertima 1986.

8. A more complete discussion of Leonardo's transformation of vision can be found in chapter 1 of my *Postphenomenology* (Ihde 1993).

9. I have traced some interesting cross-cultural aspects of the imaging of others in pre-compared to postphotographic contexts; see chapter 4 of *Postphenomenology* (Ihde 1993).

10. These patterns are used by speech pathologists, for example, to show speakers how what they are saying does not, in fact, correspond to the standard form of a native language.

11. Kevles provides a time chart, paralleling the various developments in the multiple imaging instrumentation.

12. The visit to the Cholula pyramid and conversations with anthropologists occurred during the 9th International Conference of the Society for Philosophy and Technology, November 1996.

13. Postmodernism is more thoroughly discussed in *Postphenomenology* (Ihde 1993).

14. I refer to Hacking's "if you can spray them then they are real" in *Representing and Intervening* (1983, 23).

15. This rhetoric is an example of the more-than-neutral language often employed by science reporting.

Index